Teaching Earth Science
with Investigations
and Behavioral Objectives

Teaching Earth Science
with Investigations
and Behavioral Objectives

Donna E. DeSeyn

Parker Publishing Company, Inc. West Nyack, N.Y

Library of Congress Cataloging in Publication Data

DeSeyn, Donna E
   Teaching earth science with investigations and
behavioral objectives.

   1. Earth sciences--Study and teaching (Elementary)
I. Title.
QE40.D47   372.3'57   73-12633
ISBN 0-13-892471-6

Printed in the United States of America

Dedication

To family, friends and colleagues who have
supported my efforts.

How This Book Will Help You to Teach the Earth Sciences More Effectively

Children are interested in their surroundings — their home, their school, their neighborhood. The urban, the suburban, and the rural environment each provides constant sources of investigation for the children living there. They investigate the ground as they dig a hole in a sand box, a neighbor's lawn, or a crack in the sidewalk. They open their mouths and let the rain fall on their tongues to see what it tastes like. They gaze at the clouds, watching castles and other forms pass overhead. And, they observe changes in the moon or count the stars in the night sky. This natural interest in the environment, on the part of children, is the subject of the investigations which have been compiled in the material that follows.

The investigations included in this practical book were prepared for teachers of the intermediate and middle school grades (4-9). There are many options open to you, the teacher, in how best to use these investigations in your situation. You may select investigations on the basis of the behavioral objective. The objectives may supplement previously identified objects or they may be the first effort in setting observable goals for students. You may base your selections on the subject matter content of the investigation or given experience. You may use the investigation with an entire class or select it on the basis of individual students or small groups of students. Hopefully, the wide variety of experiences contained herein will serve some of your needs and help you to develop new ideas of your own for use in your teaching.

Each chapter will provide you with a diversity of experiences to meet the range of abilities and interests of the individual students in your classroom. Some of these students may need basic introductory experiences in the processes of science. Other students may benefit from enrichment experiences, while you seek to extend the learning opportunities for higher ability students. As a result, the experiences may be conducted during class, as independent study, or as experiences to be carried on at home. In determining the objectives and the how and where of the experience, make an effort to capitalize on local conditions to add relevancy.

Part of any investigation is to identify the conditions which exist at its beginning. For the child in the middle grades, this means involvement in identifying the components of existing situations, including shape, size, color, location, and time. The components of our environment change and these changes are the object of many of the investigations which follow. These investigations provide experiences in observing, identifying, naming, measuring and recording data which can be used with your students. These experiences are important, for they provide the student both with knowledge of his surroundings and the ability to use the learning skills necessary to acquire additional knowledge. Use of these skills enables the child to identify things as they are and to establish reference points against which changes in the surroundings can be verified.

These experiences should help children recognize the constant changes taking place in their surroundings, the influence that these changes have on them, and the influence they have on these changes. The investigations included here are an effort to accomplish this through the earth sciences. Each investigation was prepared to be done by students and to present accurately and concisely the information needed by the student and teacher to permit the student to do the investigations. The content areas and objectives are identified at the beginning of the investigation. The procedures to be followed by the student include both directions and questions. A section entitled "Procedural Information" provides additional background and suggestions which you may need in working with students. Materials included in the investigations were selected on the basis of appropriateness, ease of access to the materials by the teacher, and ease of use by the student.

Diagrams and record forms have been included and can be duplicated by the teacher for student use. Other diagrams are meant to suggest to the teacher the type of materials that might be obtained from books, magazines or other sources for student use or reference. Other information provided includes sources of materials, vocabulary, precautions, and other tips for teachers.

The behaviorally-oriented investigations which are presented were selected to emphasize the endless variety of changes constantly taking place in the universe. Change is very evident in the earth sciences. It is not, however, restricted to any one discipline and for this reason the teacher can relate these investigations to other activities and areas of the curriculum. The relationship between the experiences, the student's world, and other areas of the curriculum are included as a part of the procedural information in cases where such may not be obvious. The teacher should assist the student in recognizing the relevancy of each experience.

The behavioral objectives cited here are those which define the primary goals of the experiences. Each experience can be used to achieve not only the student objective expressed but many supplementary objectives. The teacher can best identify these supplementary objectives by knowing the individual children in each class.

To aid the teacher in the evaluation process, the actual skill or knowledge the student should acquire is stated in the behavioral objective. The evaluation activity then involves the student in demonstrating the skill or knowledge stated in the objective. The level of expectation for the students is left to the discretion of the teacher.

Donna E. DeSeyn

Acknowledgments

Many of the ideas for the investigations included in this book are contributed by Richard Adamus, a teacher in the Rochester City Schools, Rochester, N.Y.; Robert Ridky, Department of Science Teaching, Syracuse University, Syracuse, N.Y.; Richard Starr, a teacher at Penfield Central School, Penfield, N.Y.; and Daniel Twigg, a teacher at the Rush-Henrietta Central School District, Henrietta, N.Y. I would like to thank this outstanding group and Sister Mary Edward Connelly of Mercy High School, Rochester, N.Y., for the part they played in developing this manuscript.

Contents

11

1

How to Observe Changes
in Shapes and Sizes

The skill of making observations is one of the most basic skills in science. By the time students reach the intermediate grades, they should be aware of the importance of using as many of the senses as possible to make observations.

Students operating at a higher level of observing will include in their observations statements of direct or indirect measurements and changes that occur during the period of observation. Making observations which include measurements may become quite challenging to students, as they must devise methods for making some measurements. Observing changes requires that the student first make a set of observations, then add necessary alterations to account for the changes which are observed over a given time period.

Making observations is just the beginning for the student. Observations can be placed in sequences or indexed. Inferences and predictions may be made on the basis of observations. All of these help develop skills which are important to the child in any subject area. Utilizing the environment as the source of observations provides both an inexpensive and endless source of interesting materials and events.

Experience 1: MEASURING THE DIAMETER OF RAINDROPS

OBJECTIVE

Measure and record the diameter of raindrops and infer the raindrop diameter.

Drop	Diameter of drop mark (mm.)
1	
2	

Diagram 1

Drop	Diameter of drop (mm.)	Diameter of drop mark (mm.)
1		
2		

Diagram 2

PROCEDURE

a. Mount one thickness of nylon stocking tightly on the mouth of a jar, or on embroidery hoops, and coat the nylon with powdered confectioners' sugar. Shake off the excess sugar.

b. Using a medicine dropper to simulate a raindrop, let a drop fall onto the powdered screen and describe what you observe.

c. Measure the diameter of the drop marks on the nylon. Repeat this procedure and record the measurements for at least five drops made with the medicine dropper. See Diagram 1.

d. On the basis of your observations, are the drop marks the same size? What is the range in size of the drops?

e. Are the drop marks on the nylon a good indicator of the actual falling drop size? Does a mark of 2 mm. in diameter on the nylon mean the mark was made by a 2 mm. drop? Measure the diameter of at least 10 different drops hanging from the medicine dropper and record the measurements. Compare the drop size with the size of the mark made on the nylon. See Diagram 2.

f. Graph your results from e., plotting the actual drop diameter on the y axis and the drop mark diameter on the x axis.

g. What is the maximum size drop you were successful in producing? The smallest?

h. The next time it rains, take several powdered nylon *raindrop recorders* outside. Briefly expose a different recorder to the rain at one minute intervals and answer the following questions based on your observations and measurements:

Are the drop marks of equal size on the first screen?

How do they compare with the later drop sizes?

What is the largest drop recorded? The smallest?

Describe any change in the drop size as the rain progresses.

What drop size occurs most frequently? Would the same results be true of all storms?

What is the range of the drop size?

Does the height from which the drop falls affect the mark size? (Try varying the height from which drops of the same size are released.)

MATERIALS

used nylon stocking (60 gauge)
confectioners' sugar
wide-mouth jar with canning ring, rubber band
or embroidery hoops
medicine dropper
metric ruler

PROCEDURAL INFORMATION

The students can probably recall that there were times when the size of the raindrops must have been larger than they were at other times. The very idea that it is possible for them to measure the size of raindrops will be intriguing for many students. Some may even want to extend this experience by estimating the volume of their raindrops and the amount of rain that falls during a storm. This type of activity involves the child in developing a technique for measuring observable differences in his environment.

Women's nylon stockings (60 gauge), covered with confectioners' powdered sugar, function very nicely in recording raindrop size. Many other types of apparatus have been used, but sugar-covered nylon is simple, inexpensive, and gives satisfactory results.

The nylon should be tightly mounted on a frame Embroidery hoops or a mason jar with a canning ring or rubber band will serve as frames to hold the nylon. Once mounted, the nylon is coated with sifted confectioners' sugar until it looks powdery white.

Before actually measuring raindrops, students will practice with "drops" which they manufacture. Medicine droppers, squirt bottles, sections of hoses of various diameters, straws, and anything else that could produce drops are satisfactory. Metric rulers can be used to measure the diameter of drops made by various means. The actual drop diameter should be compared with the drop marks made by the drop as it hits the powdered nylon. A graph of the results will give a visual image of what is happening.

In simulating the raindrop, circles should be observed on the powdered screen. As the drop hits the powdered nylon, it moves the sugar through the nylon, leaving the circle sugarless. The circles formed by a given drop size will be fairly uniform in size, with different size drops producing different size circles or marks. The marks serve as valid indicators of the drop size although the marks are generally larger than the corresponding drop. The maximum size of natural or artificial raindrops would be about 6 mm., while the smallest ones would vary from 1-2 mm. in diameter. Sometimes, the winds sort raindrops by size as the smaller drops are moved farther and fall at the same time. See Diagram 3.

It is important that the nylon used must be a worn one rather than a freshly washed one, as the body oils which accumulate on the stocking appear to allow the sugar to stick to the nylon.

EVALUATION

During a rainstorm, the drops made marks which were recorded as 8-10 mm. in diameter: infer the size of the raindrops.

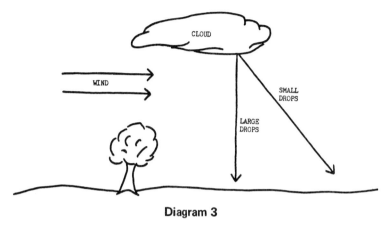

Diagram 3

Experience 2: OBSERVING CHANGES IN A PUDDLE

OBJECTIVES

a. Observe and measure changes of shape and size which occur in a puddle.

b. Infer the causes of these changes and predict future changes.

PROCEDURE

a. After a rainstorm, go outside and find a puddle that is easy to measure (about 1 to 3 feet in diameter).

b. Outline the puddle with a chalk mark or a piece of string.

c. Measure the widest part of the puddle, marking both spots where the measurement was made by small sticks or stones. These will serve as reference marks for future measurements.

Record this measurement and the time it was made.

d. After 15 minutes repeat step c. Make a third measurement after another 15 minutes.

e. Answer the following based on your observations:

Was there a change in the shape of the puddle?
If so, describe the change.
Was there a change in the width of the puddle?
If so, describe the change.
Predict how long it would be before the puddle would disappear.
What factors control the disappearance of the puddle? Explain.

VOCABULARY

evaporation   absorption

MATERIALS

chalk or string   thermometer (optional)
a watch   hygrometer (optional)
meter stick   anemometer (optional)
                 small sticks or stones

PROCEDURAL INFORMATION

The nature of the surface on which the puddle is found will influence the rate at which the puddle disappears. A marked difference may be observed if one group selects a paved surface and another a dirt surface. Different dirt surfaces will, also, affect the rate.

Predictions may consider the nature of the surface, the temperature, wind and relative humidity. The more sophisticated these observations are, the more accurate the predictions should be. If there is a lake or pond in the vicinity, the students may extend their inferences and predictions from the puddle to these larger bodies of water.

EVALUATION

Given 2 drops of water, each on a separate piece of aluminum foil, one kept by an open window and the other in the middle of the classroom, predict which will disappear first and tell why you think so.

Experience 3: INDEXING CHANGES IN SHAPE AND SIZE BY TIME

OBJECTIVE

Index pictures on the basis of changes of shape and size of an object which occur over a period of time.

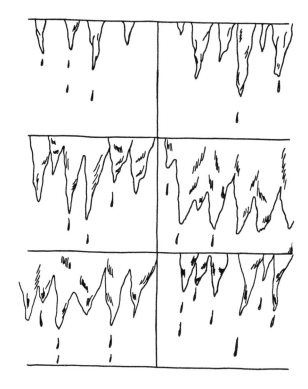

Diagram 4

PROCEDURE

a. Arrange the accompanying pictures so they show a proper sequence of changes. See Diagram 4.

b. Compare your order of changes with that of others in your class and discuss the reasons for your decisions.

c. Cut pictures from a magazine or draw pictures to show changes in shape and/or size which occur in nature over short and long periods of time. Exchange picture sets and see if someone else would sequence them as you did.

VOCABULARY

index   sequence

MATERIALS

1 copy per student of sequence in step a.
1 copy per student of evaluation sequence
magazines (for cutting)
scissors

PROCEDURAL INFORMATION

The purpose of the experience is to have the student consider the factors which change and that serve as visual clues which can be used to sequence naturally occuring changes. This can be done by asking the student what factors he considered in deciding how to arrange the pictures. In the case of the pictures in step a, a controversy may occur over whether the pictures are of icicles or stalactites. If they are icicles, are they just forming or are they melting? The student's explanation of his sequence may make his sequence acceptable even though it is not the same as everyone else's.

Any set of pictures showing sequential changes may be used. Student drawings, cartoons or clippings from magazines may produce some excellent sequences. In discussing any of the picture sequences, it should be brought out that we predict changes on the basis of our past experiences or acquired knowledge. There are unexpected events and differences in cultural backgrounds which alter anticipated patterns and often pose new problems to scientists.

EVALUATION

Given a set of photographs showing changes in the shape (or size) of an object or region, the student will arrange them in acceptable sequence based on time.

Experience 4: DISTINGUISH THE INFLUENCE OF POSITION ON SIZE

OBJECTIVE

Distinguish the relative sizes of familiar objects in pictures, infer causes of differences and state the relationship between size and distance.

Diagram 5

PROCEDURE

a. Observe the diagram of the railroad tracks and their surroundings. See Diagram 5.

b. What is the relative size of the ties in the bottom of the diagram compared to those in the center of the diagram?

c. What is the actual size of both sets of ties?

d. How do you explain the relative difference?

e. Assuming that the tracks are four feet apart in the foreground, how far apart do they appear to be at the other end? How far apart are they actually?

f. Observe the weed to the left of the track in the foreground and compare it with the tree to the right of the track in the center of the diagram. Which is apparently larger? Which is actually larger?

g. Examine other photos and compare objects in them for apparent size at different distances as you have here.

h. State your conclusion as to the relationship between relative size and distance.

VOCABULARY

relative size

MATERIALS

a diagram or picture similar to that of the railroad tracks or the actual outdoor setting

a picture showing differences in apparent size such as buildings, trees, fences, sidewalks or streets

PROCEDURAL INFORMATION

The investigation should make clear to the student the need to know the position of the viewer in order to understand size relationships in pictures. Such pictures can be compared to maps or drawings made to scale. Map keys identify the relationship between measurements made on the map and on the actual map site. Pictures do not provide keys, so the viewer must be more alert to possible distortions introduced in a photograph. For this reason, pictures cannot be used for accurate representation unless careful attention is given to the distortion resulting in the distance from the observer to the object.

Observations of the diagram will reveal that the ties at the bottom of the diagram appear much larger even though, in reality, all the ties are the same length. Similarly, the distant ties appear to be very close while all the ties are actually 4' apart.

The relative size is dependent upon the relative distance from the object.

EVALUATION

Given a picture showing differences in the apparent size of objects, the student will:

a. Compare the apparent size of objects in the foreground and similar type objects in the background.

b. State an inference explaining the difference between apparent size and actual size.

c. State a relationship between relative size and the distance from the observer.

Experience 5: PREDICT THE RELATIONSHIP BETWEEN THE DIAMETER AND CIRCUMFERENCE OF CIRCULAR OBJECTS

OBJECTIVE

Predict the value for π (pi).

Object Name	Diameter	Circumference	

Diagram 6

PROCEDURE

a. Find 10 circular objects such as cans, cups, jar tops, water pipes, chair legs and coins. Measure the circumference and diameter of each of the objects. Record your data on the chart. See Diagram 6.

b. Label the top of the fourth column **c/d**. For each object measured, divide the circumference by the diameter and record the answer in the fourth column.

c. Are the values in the fourth column similar or different? Explain.

d. Compare your results with those of other students. Did they get results similar to yours?

e. What rule can you make about the relationship between a circle's circumference and its diameter? This value you've identified is called pi, symbolized as π. Pi is a constant when comparing a circle's diameter and circumference.

VOCABULARY

diameter   circumference

             pi

MATERIALS

rulers   variety of circular objects

        string

PROCEDURAL INFORMATION

When the students locate and measure the diameter and circumference of 10 circular objects, there should be a wide range of sizes and objects available. The students may gather the objects to be used. The results of their measurements and division, depending on their accuracy, should be similar. They should find pi to be 3.0-3.2. Pi has been measured to be 3.14159265 –.

The students should find that the values for c/d in the fourth column are similar. Their rule should indicate that the ratio between a circle's circumference and the diameter is the same regardless of the size of the circle. It would be advisable to indicate the variety of instances when π is used, such as in finding the circumference of a circle ($C=2\pi r$; $C=\pi D$), or the volume of a circle ($V=4/3\pi D$).

As a note of caution when measuring diameters, it is essential that the measurements be made through the circle's center.

String can be wrapped around the object and then placed next to a ruler for determining circumference.

EVALUATION

Given a circular auditorium 628′ in circumference and 200′ in diameter, what is the value of π for the auditorium?

2

Identification of the Influence of Particle Sizes and Shapes on Soils

Traveling through various parts of the United States, one becomes aware of the very noticeable differences in soil. There are red clays, tan loams, black mucklands, and sands of many shades. Just as the soils vary in color, so do they vary in origin, use and composition.

A quick observation of soil samples gathered from the same neighborhood might lead one to the hasty conclusion that the samples are all alike. That is possible. However, it is also very possible that they are each very different from the other. These differences are observable and measurable and form the basis for the following series of experiences.

It is highly unlikely that students would complete each of the experiences in this series. Rather, the teacher or students might select only those experiences which provide the desired skills or knowledge. The series demonstrates how an easily available material can be used in a variety of ways to provide students with meaningful experiences which are process oriented and which have content implications.

Experience 1: OBSERVING TEXTURE UNDER MAGNIFICATION

OBJECTIVE

Observe and describe visual characteristics of granular objects as viewed through magnifiers with increasing magnification.

CHART FOR OBSERVATIONS

	SALT		SAND		SOIL	
	Picture	Description	Picture	Description	Picture	Description
Naked eye						
Hand lens						
10X Tripod Magnifier						
50X Microscope						
Tactile (optional)						

Diagram 7

PROCEDURE

a. Place a pinch of salt on a piece of paper. Observe the color and shape of the salt particles with the naked eye. Place the metric ruler next to the salt particles. Make a simple diagram of what is observed and describe your observation in a short sentence on the accompanying chart. See Diagram 7.

b. With a hand lens (less than 10x magnification), view the salt again. Describe what you see by a drawing and a few words on the chart.

c. Using a 10x tripod magnifier, view the salt and again record your observations on the chart.

d. Repeat the process, using 50x (or more) microscope, and record what you observe on the chart.

e. Repeat the entire process (a-d), using first a pinch of sand and then a pinch of soil. Record and compare your observations in each case.

VOCABULARY

magnify   microscope
tripod   hand lens
10x   crystals

MATERIALS

hand lens   sand
tripod magnifier   metric ruler
salt   observation charts
                    microscope (50x or greater)

PROCEDURAL INFORMATION

The students will have several occasions to examine particle sizes in future experiences. This is an opportunity for the

Diagram 8

teacher to be sure all students are adept at using the various instruments available to them. This experience, also, can be used to discuss the importance of both the visual and tactile senses in making observations by adding another column on the chart for tactile observations.

EVALUATION

Draw a line from the picture on the left to the picture on the right that best shows how salt would appear when using the tool in the left column. See Diagram 8.

Experience 2: MEASURING SOIL PARTICLE SIZES AND SHAPES

OBJECTIVE

Measure and record the size and shape of particles found in various soil samples.

SOURCE OF SOIL SAMPLE	TOTAL VOLUME BEFORE SCREENING	VERY FINE Vol. %	FINE Vol %	MEDIUM Vol %	COARSE Vol %	VERY COARSE Vol %

Diagram 9

PROCEDURE

a. Obtain at least 3-4 samples which can be used by the entire class. The samples will be used in several investigations so at least a pail-full of each sample is necessary. The materials might be obtained from: (1) field or vacant lot, (2) sand bank, (3) gravel pit, (4) beach, (5) stream, (6) construction site, (7) backyard, (8) school grounds. Label each sample by its source.

b. Select a soil sample for your group to use. Starting with a known amount of your sample, for example one pint jar full, sort the sample on the basis of particle size using the screen sieves. As each sample is sorted, place the sieved, graded soil samples into separate labeled piles such as very fine, fine, medium, coarse, and very coarse.

c. After obtaining graded samples (4-6 depending on the number of screen sizes available) of each material collected, estimate what part of the original sample was very fine, fine, medium, coarse, and very coarse material. Record the estimates as fractions or percents on the record sheet. The estimates may be based on relative mass or volume. See Diagram 9.

d. Examine each of the graded samples. Are there any similarities between the particles in a pile other than size? Use a magnifying glass or hand lens to examine the sample labeled "very fine." Do you think these materials could be separated again? Explain.

e. Use your ruler to determine the average size of the particles in the graded samples and record your measurements. Measuring such small particles can be made easier by placing a piece of clear acetate on the overhead projector. A sample of the smallest particles can be sprinkled on the acetate and projected on the wall. Placing a transparent metric ruler on the acetate next to the particles, projects the ruler onto the wall or screen so that the small distance of 1 mm. appears enlarged to several centimeters, and you can subdivide this with ease. By holding a piece of paper on the wall where the projected image is seen, both the ruler markings and the particle outlines can be traced. From these, the particle size can be determined.

f. While the particles are still projected, describe their general shape and compare the shape of these particles with the shape of the very coarse particles.

g. List the ways in which each of the original samples (from a) are alike. List the differences between each of the original samples on the basis of observations.

h. Place the graded samples in the labeled containers and save them for the following experiences.

MATERIALS

several samples of earth from different sources
transparent plastic, metric ruler
several transparent glass or plastic jars with labels
overhead projector
hand lenses
4-6 different sizes of sieves, commercially available screen sieve sets (Hubbard Scientific) or sieves can be made by purchasing different sizes of metal screening at a hardware store, removing part of the bottoms of several shoe boxes and covering each hole with a piece of screening.

PROCEDURAL INFORMATION

You may prefer to have each student perform the entire experience or you may prefer to have small groups of students using single samples. The latter would mean that the chart used would be completed as a class record. If done this way, be sure that the students examine the sieved materials from other groups.

The purpose of the experience is to show that earth materials are composed of different sized particles, and the quantity of any given size varies from sample to sample. The processes that form soils usually tend to produce rounded fragments as opposed to angular ones.

A problem may be encountered in attempting to directly measure the smallest particles. A projected image easily solves the problem. By projecting a transparent ruler even the smallest divisions can be further subdivided.

Be sure each student carefully labels the 4-6 graded soil sample containers as the soil samples are needed for other experiences in this investigation. Save enough unsieved soil to be used in Experience 3.

EVALUATION

Given a sample of unknown earth material and the equipment used in Experience 1, determine the average size and most common shape of the particles making up the sample.

Experience 3: SEPARATING FINELY GRADED SOIL SAMPLES

OBJECTIVE

Observe and identify the particles making up the "very fine" earth material.

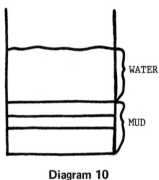

Diagram 10

PROCEDURE

a. When you sieved the earth materials there was some material which passed through even the finest screen sieve. Does this indicate that this very fine material is all the same size? (Refer to the tracing made when measuring the very fine particles.) Explain your answer.

b. If it is desirable to try to further separate the very fine material, how would you go about it? Can you think of more than one way to do it?

c. Place about 4-5 cm. of the very fine material in a large test tube or cylinder and add water until the tube is nearly full. Shake the tube vigorously to mix the contents thoroughly. Set the tube down and describe what you see.

d. Observe the tube for five minutes, describing any change(s) that you see occurring. Describe the tube's appearance at the beginning and at the end of the five-minute period.

e. You may have found that some material settled almost immediately. Would this be the largest or smallest particles of the very fine material?

f. Was the water still muddy after five minutes? What size material would still be suspended in water after this time?

g. Set the tube aside and check it each day until the water is clear. Keep a daily record of your observations.

h. When the water has cleared, make a drawing of the material that has settled and describe any layers by color and depth.

i. Carefully pour off the water and spread the material on a tray to dry. Replace it in its container for further use in following experiences when it is dry.

MATERIALS

graduated cylinder or other tall, transparent container
very fine graded soil sample (from experience 2)
water   timer   ruler

PROCEDURAL INFORMATION

Precautions should be taken to avoid spillage while shaking the water and soil in the cylinder. A covered container or a large cork for the cylinder might prevent an accident.

Upon shaking, the water becomes "muddy" in appearance. The majority of the material can be observed accumulating at the bottom when shaking ceases. The water still remains murky, however, indicating differences in particle sizes in the sample. Some of this very fine material could remain in suspension for several days. The suspended material is clay and it can be more easily observed by shining a bright light into the container.

Be sure to dry and save the soil samples used in this experience. They should be returned to the labeled containers for later use.

EVALUATIONS

a. A student went to a muddy river and scooped out a

container of the water. In one week, the container had layers of "mud" on the bottom that resembled the accompanying sketch. Describe the particle sizes that probably compose the "mud" layers. See Diagram 10.

b. Many streams, even large rivers, dry up on occasion. While visiting a dried-up river channel, an observer noted that an extremely fine dust was on top of the surface, with larger and larger particles encountered as he dug into the channel. Explain these observations.

Experience 4: INDEXING SOIL SAMPLES ON THE BASIS OF OBSERVATIONS

OBJECTIVE

Describe various soils and index them on the basis of observations.

Sample source	Description	Description when wet	Description when dry

Diagram 11

PROCEDURE

a. Feel each soil sample (unsieved original samples saved from Experience 2) and describe its characteristics on the accompanying chart. See Diagram 11.

b. Using the sieved soil samples from Experience 2 which were obtained from the same source as just used, carefully examine them by rubbing some from each graded sample, both

Color of Sample

Original S.	V.F.	F.	M.	C.	V.C.

Diagram 12

wet and dry, between two fingers. Record your observations on the accompanying chart.

c. Do materials act the same when dry as when wet? Explain.

d. Describe the color of the total soil sample and then the colors of the soil's components as sorted by particle size. Record your observations on the accompanying chart. See Diagram 12.

e. Using any or all of the information you have accumulated, index the soil samples collected into a classification scheme.

f. Dry all the graded soil samples and return them to the proper labeled tubes for the next experience.

MATERIALS

sieved and unsieved soil samples from Experience 2
water
unknown soil sample and sieves (for evaluation)

PROCEDURAL INFORMATION

It may be that additional materials will have to be screened or re-sieved. When comparing the wet and dry samples by rubbing, it may be surprising to find that the smaller particles feel powdery when dry but slippery when wet. This is often due to the presence of clay particles.

Again all soil samples should be dried and saved for other experiences.

EVALUATION

Given an unidentified soil sample describe it and index it on the basis of information recorded on your chart.

Experience 5: INFER THE EFFECT OF DIFFERENT SOILS ON ABSORPTION

OBJECTIVE

Measure the rate at which water is absorbed by soil materials and graph the results. Infer the effect of different types of soils on absorption of rain water.

Particle size of sample	Absorption time (sec.)	Distance	Rate of Absorption (mm./sec.)

Diagram 13

PROCEDURE

a. Place 8-10 cm. of graded soil samples from Experience 2 (very fine, fine, medium, coarse and very coarse) in similar graduated cylinders, one sample per cylinder.

b. Pour equal amounts of water into the top of each cylinder and measure the time it takes for the water to reach the bottom of each container. Record your findings on the chart for each sample. See Diagram 13.

c. Make a graph of your results, using particle diameter (data from Experience 2) on one axis and the time required for the water to travel the length of the cylinder on the other axis.

d. Soil and earth materials vary in their particle size. Study the graph you have made. What happens to the rate of absorption when the diameter of the material increases? Does the absorption rate double if the diameter doubles? What can you infer about the relationship between particle size and the rate of absorption? What size particles do you think would allow for the greatest amount of absorption after a rain? The least? Explain your answers.

e. Dry and save the soil samples for use in following experiences.

VOCABULARY

absorption rate

MATERIALS

4-6 graded soil samples from Experience 2
water
timer
graduated cylinders (or other tall transparent containers of the same size)
transparent plastic ruler

PROCEDURAL INFORMATION

The students time how long it takes the water to soak into various-sized soil samples. The results are graphed to see the effect of particle size on the water absorption rate. The actual data collected will vary with the particle sizes used. The amount of water used will depend upon the size of the containers used.

Once completed, the graph can be used for predictions. For example, what would be the absorption rate for a material composed of 3½ mm. diameter particles (assuming 3½ mm. particles had not been used).

EVALUATION

Given some sieved material, infer its absorption rate and check your inference.

Experience 6: DESCRIBE THE RELATIONSHIP BETWEEN CAPILLARITY AND PARTICLE SIZE

OBJECTIVE

Measure the capillarity of graded soil samples and describe the relationship between capillarity and particle size.

PROCEDURE

a. Place each of the graded soil samples (from Experience 2) into separate plastic tubes which have a piece of cloth fastened to one end to contain the sample. Fill the tubes nearly full.

b. Using the very coarse sample, place the cloth end of the tube into a paper cup of water so the end is just submerged. Make a chart and show the different times it takes the water to rise one centimeter in the tube.

c. Describe what you observe happening in the tube. Does the water advance upward at a uniform rate or does it move in spurts?

d. Repeat this procedure, using another graded soil sample until the range of samples from very coarse to very fine has been used.

e. Summarize your findings from the chart in a few sentences.

f. Explain why a lawn planted in a fine soil tends to stay greener longer than a similar lawn planted in coarse soil.

VOCABULARY

capillarity   capillary action
            water table

MATERIALS

plastic tubes (these can be made by rolling a sheet of transparency acetate into a tube about 2" in diameter and stapling)
graded soil samples (from Experience 2)
pieces of cheesecloth or cotton

string or tape   water
paper cups   timer
                ruler

PROCEDURAL INFORMATION

The students may have observed the creeping upward of water in a piece of cloth or paper or they may have heard that something similar happens in plants. They may not be able to identify that it is capillary action which causes this motion. A discussion or simple demonstration of what happens when one corner of a piece of cloth is placed in water may help establish a common ground for the students before starting the experience. Capillary action in soils provides students with a hands-on experience as they observe and measure the effect of particle sizes on the moisture in a soil.

Capillary action will cause water to rise in a soil sample primarily due to molecular attraction. When large-particle sizes are used, however, the spaces between particles can become so large that intermolecular forces are not sufficiently strong to "lift" the water. Thus we find that a fine soil will have water coming "up through" as a result of capillary action. This rising water can sustain plants over a dry period. The coarse soil, on the other hand, does not experience as much capillary motion and plants must therefore exist on water contained in the root zone.

EVALUATION

Given a very coarse sample and a finely-graded sample of soil, tell which sample would demonstrate the greater capillarity and explain your choice.

Experience 7: MEASURING THE DRYING RATE OF SOILS

OBJECTIVE

Measure the drying time for wet earth materials and infer the effect on gardens and lawns.

PROCEDURE

a. Earlier, you completed an experience in which you measured the rate at which water was absorbed by graded soil samples. Review the findings from that activity.

b. Place equal amounts of graded soil samples prepared in Experience 2 into plastic containers. A depth of 4-5 cm. would be adequate. Add an equal volume of water to each sample, being sure to dampen the samples thoroughly. Note the appearance of each sample and record the time and date the sample was prepared. Check the samples periodically until the soil is dry. Record your observations of the time it took for each sample to dry.

c. What do you think would have happened if the material had been put in some other container, such as a tube having less surface area of the sample exposed to the air?

d. If your father's garden or mother's flower garden is composed primarily of rather small particles, should they water it more or less frequently than if it was composed of larger particles of soil? Explain.

e. Save the dry graded soil samples for the next experience.

MATERIALS

  plastic shoe boxes or trays spoon
  graded soil samples   water

PROCEDURAL INFORMATION

Students have seen that water is absorbed at different rates by different materials. Is the reverse true? Does wet soil dry out at different rates? This experience helps them find out by trying it in simulated conditions. Placing the samples in sunshine will accelerate results. The rate of evaporation will be much slower than the rate of absorption. All samples should be in similar conditions to facilitate accurate measurements. The fine material will dry more rapidly than the coarse. This is primarily due to capillarity.

EVALUATION

Your lawn is planted in a very fine-sized dirt. Is there danger of the lawn drying out? Explain.

Experience 8: DESCRIBE THE EFFECT OF SOIL PARTICLE SIZE ON PLANT GROWTH

OBJECTIVE

Measure the growth rate of plants in sieved soil materials and describe the results.

PROCEDURE

a. Use a cupful of each of the sieved graded soil samples from Experience 2 and a cupful of each of the original sieved samples. The latter will provide a comparison.

b. Place 3-5 soaked bean seeds in each container, being sure to place all seeds at a similar depth. Give each cup a predetermined volume of water and place the cups on a window sill or table so they will receive sunlight.

c. Observe the cups of seeds daily, watering them and recording the data on the accompanying chart. See Diagram 14.

d. Which soil sample seemed best for growing the plants? Which sample of sieved soil seemed best for growing the plants? Which soil sample was least successful? Which sieved sample was least successful? To what do you attribute your results?

MATERIALS

original soil samples from Experience 2

sieved samples from Experience 2
bean seeds
plastic cups, 6 per group
water

Very coarse	No. Days Growing									
	Avg. Height									
	Avg. No. Leaves									
	Avg. Diameter									
	% Increase in Hgt.									
Coarse	No. Days Growing									
	Avg. Height									
	Avg. No. Leaves									
	Avg. Diameter									
	% Increase in Hgt.									
Medium	No. Days Growing									
	Avg. Height									
	Avg. No. Leaves									
	Avg. Diameter									
	% Increase in Hgt.									
Fine	No. Days Growing									
	Avg. Height									
	Avg. No. Leaves									
	Avg. Diameter									
	% Increase in Hgt.									
Very fine	No. Days Growing									
	Avg. Height									
	Avg. No. Leaves									
	Avg. Diameter									
	% Increase in Hgt.									
	Length of growing period									

Number planted _____
Seed used _____
Date planted _____

Diagram 14

PROCEDURAL INFORMATION

The students should plan and provide for the daily care of their plants, as well as for care over weekends and vacations. Soaking the seeds overnight before planting will speed up germination. Also, using mung beans will shorten the required growing time considerably. Mung beans are available from health food stores.

The questions raised in d should result in considerable discussion as the students attempt to identify which soil has produced the best plants. They will have to determine what is meant by best and their criteria for evaluating the results of measuring their plants. They should consider if the different soils are the cause of the differences in the plants or if there were other factors involved. It should be possible for them to measure some differences in the plants and to infer some relationships between the soil and the plants.

EVALUATION

Given a choice of three soil samples: very fine, very coarse and an original ungraded soil; beans soaked for 24 hours; and a plastic cup; select one of the three soil samples to grow your seeds and give reasons for your choice.

3

How to Identify and Measure
Stream Changes from
Maps and Model Construction

Models are often an excellent instructional aid when it is not feasible to use the real object. They are also helpful in preliminary studies conducted in the classroom, prior to going to the field. Much can be learned indoors about the changes which occur in a stream through the use of models.

Investigations using stream models become meaningful when followed by similar studies of the real situations in nature. No amount of textbook reading or laboratory experiences can provide an adequate substitute for field studies. This is true whether the fields are the actual fields or the parking lots and streets. The classroom models do much to prepare the student for acquiring greater satisfaction from the field studies.

The use of the model in the classroom presents one major problem which can be overcome. When working with a model, the children become familiar with the ideal or stylized situation. The child who then goes into the field looking for an example of the model may become discouraged by being unable to find the real thing. It is then highly desirable for the children to get into the field with the teacher to find, together, less than perfect examples of their models.

Experience 1: MEASURING STREAM GRADIENT FROM MAPS

OBJECTIVE

Determine the gradient of a stream, as represented on a topographic map, by making the appropriate measurements and arithmetic calculations.

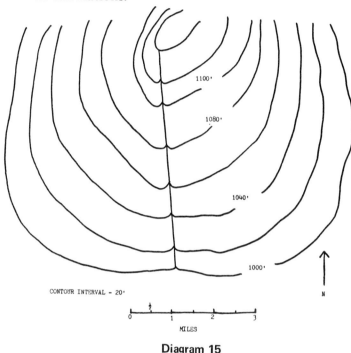

CONTOUR INTERVAL – 20'

N

MILES

Diagram 15

PROCEDURE

a. The accompanying map shows a stream flowing down a hill. A spring might be the source of the stream. How long is the stream? See Diagram 15.

b. What is the elevation of the stream's source?

c. Which direction is the stream flowing? Give two reasons why you think so.

d. From its beginning to end on the map, how many feet does the stream "drop"? Show your work.

e. You have found the length of the stream and the number of feet it drops. What is the average drop in elevation per mile for this stream? Show your work. This average drop in elevation (feet/mi.) is known as the average gradient of the stream.

VOCABULARY

topographic map   spring
source   gradient

MATERIALS

map similar to the one shown, 1 per student
evaluation map, 1 per student
ruler

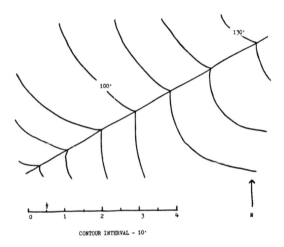

CONTOUR INTERVAL - 10'

Diagram 16

PROCEDURAL INFORMATION

Streams are seldom perfectly straight and rarely begin or end on a contour line. After using the idealized stream to introduce the process, other stream gradients can be used. The average gradient or slope of a stream is simply the average drop in elevation per mile expressed in feet per mile. This is arrived at by determining the stream's length and the drop in elevation, then dividing the elevation change by the stream's length.

Students have usually learned the phrase "water runs downhill" so well that it is necessary for them to work through an experience such as this to realize that all north-south rivers in the northern hemisphere do not flow south.

The stream shown is 5 miles long, originating at a source 1140 feet above sea level. The stream flows south, since that is the downhill direction as indicated by the contours. The contours also bend to the north, indicating the uphill direction. The water in the stream flows from 1,400' to 1,000' or a total drop of 140'. The average drop for the stream is 28' per mile.

EVALUATION

Determine the average gradient of the stream on the accompanying map. See Diagram 16.

Experience 2: IDENTIFYING COMMON STREAM CHARACTERISTICS

OBJECTIVE

Distinguish the characteristics of streams in different stages of development, describe them by simple diagrams and identify them by name.

PROCEDURE

Examine the two maps showing different sections of the Genesee River and answer the following.

a. In what direction is the river flowing?

b. Compare the path the river follows on the two map sections.

c. Stretch a piece of string between the place the river begins and where it ends on the Genesee map. The string represents a distance which is often referred to as "as the crow flies" or a straight line. How long is this distance?

d. If two towns were located on the river 20 miles apart "as the crow flies," which section of the river would provide you the shortest canoe trip between the two towns? Explain.

e. Identify the river valley on each map section and compare the river width with the valley width in each case.

f. The following are features associated with some streams or rivers. They may or may not be present on the maps you're using. Examine each section of river and make a chart showing the features found and the section of the map on which they are found. Use encyclopedias, textbooks, or other reference books to identify features you don't recognize.

oxbow lake	oxbow scar
meander	swamp
rapids	canyon or gorge
waterfall	flood plain

VOCABULARY

oxbow lake	rapids
meander	oxbow scar
swamp	flood plain

MATERIALS

topographic maps of Genesee Quadrangle, New York — Livingston County;

topographic maps of Rochester East Quadrangle, New York — Monroe County;

topographic maps of Rochester West Quadrangle, New York — Monroe County (optional); (for sale by U. S. Geological Survey, Washington 25, D. C.; a descriptive folder on topographic maps and symbols is also available upon request)

string

PROCEDURAL INFORMATION

Rivers tend to change over a period of time as erosion and deposition take place. The water in a young river tends to run swiftly in a relatively straight course, with many waterfalls and rapids. Rapid down-cutting by erosion often results in the river occupying the entire floor of a steep-walled gorge or canyon. This erosion reduces the waterfalls, rapids, and other features of the young river until the river takes on a new set of characteristics. The river then has a wide flood plain, no longer occupies the entire valley floor, has numerous meanders, and oxbow lakes and oxbow scars may be present. The water is sluggish and slow-moving, characteristic of the old stage of a river.

The Genesee River shows all of the features listed in the activity in the two sections of map used. It is an illustration of a river which has the characteristic stages of both the young river and the old river, in close proximity. The fact that the Genesee River flows north was another reason for its selection. This presents a problem for some students when looking at a wall map of the United States. North is "up" on the map and water doesn't flow uphill. Using a north-flowing river helps relate this experience with the previous one.

EVALUATION

Separate the following river features into two groups, one with fast-moving water and the other with slower-moving water. Label the fast-moving list "young" and the slow-moving list "old." Describe any four of the terms on your lists.

meanders	oxbow scar
rapids	waterfalls
oxbow lake	swamps
canyon (or gorge)	flood plain

Experience 3: CONSTRUCTING A STREAM MODEL TO OBSERVE RUNNING WATER

OBJECTIVE

Construct a stream model and observe the erosional and depositional changes which occur.

PROCEDURE

a. Fill about ½ of the bottom of the plastic tray with dirt about 3-4 inches thick. Make a simple sketch of the model.

b. Soak the dirt in your model, being careful not to move the dirt particles. When the dirt is moist, raise the dirt end of the tray one or two inches and place a book or other support under it. Pour some water into the lower end of the tray to form a lake. The land and lake should just touch.

c. Fill a plastic squeeze bottle with water and direct a flow of water from the bottle toward one area on the higher end of the land mass. Continue the flow for about 1 minute, then stop and answer these questions:

What happened as the water moved over the land?

What changes can you observe after the water ran over the land for 1 minute?

What is the size of the largest "rock" moved by the stream? The smallest sized particles?

Has a stream been established?

What changes have occurred in the water by the time it reaches the lake?

d. Make a sketch of your model.

e. Squirt the water for about three minutes more on the same area before answering the following questions.

What additional changes can you observe in your model?

Where did the material go that was carried by the stream?

Draw a sketch of the top view of the formation that is building up at the mouth of the stream. The water-formed feature is called a delta. Describe how the delta formed.

As more water was squirted into the model, what happened to the level of the lake? Why doesn't the same thing occur in real lakes?

f. Sketch the gorge formed by the river. What is the depth of the gorge? The length? The width? Using this data, determine the volume of the gorge (your teacher will assist you).

g. Carefully remove the delta material with a spoon and put it into the graduated cylinder. What is its volume? Compare the volume of the gorge with the volume of the delta material.

h. Based on your observations of the model, what size particles would you expect to find in a delta? Explain.

VOCABULARY

canyon   delta   gorge

MATERIALS

plastic box, tray or dishpan (box lined with aluminum foil)
plastic squeeze bottle or empty liquid detergent bottle
dirt mixture with fine-to-coarse materials
ruler
graduated cylinder
plastic spoon

PROCEDURAL INFORMATION

There are many observations that can be made while observing a model stream and some students may wish to repeat the activity introducing variations. If the dirt used is from a driveway or roadbed, it may contain a variety of foreign objects from twigs to feathers. A few expected observations could include: the stream moves materials by carrying, rolling or bouncing them along; larger "rocks" are uncovered that can't be moved; bends in the channel develop near uncovered "rocks;" once started, a meander continues as the stream cuts on the outside of the bend; a gully develops where material is removed; a gully resembles a real gully with larger unmovable objects such as logs (twigs) and stones causing rapids; transported material is deposited in the lake as a fan-shaped delta with the larger particles deposited first and finer particles deposited last; the stream may change its course on the delta through minor streams or distributaries.

The relationships of the influence of natural resources, such as rivers, on the people in an area can be emphasized if both the earth science teacher and social studies teacher can combine their efforts. Early settlers were very aware of the advantages of settling on or near a river. The impact of the changes of a river, its flooding, changing course, and even its pollution cut across the boundaries of academic disciplines.

If possible, take the students to observe a real stream so they can compare it with their model. If a stream is not available, perhaps temporary streams could be studied on the school grounds after a rain.

PRECAUTIONS

Try this activity to familiarize yourself in advance with what the students will be expected to see. You can observe the need for soaking the dirt before starting the stream of water from the squeeze bottle. To avoid water spillage, check the plastic boxes to be sure they won't leak. Also, it may be necessary to siphon off some of the lake water to prevent it from getting too deep to form a good delta.

Washed sand, in which most of the particles are more rounded and even-sized, gives different results from a gravel mixture. The sand is more apt to collapse and shift rather than hold together on the sides of gorges. Students using a variety of materials may well infer that the nature of the soils and rocks influences whether a river develops gorges or canyons.

EVALUATION

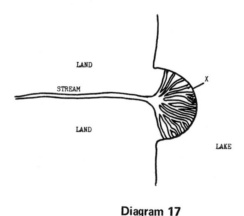

Diagram 17

Using the accompanying diagram (Diagram 17):

a. Identify the feature labeled X and describe how it was formed.

b. The stream is probably in a valley or gorge. What reason can you give for this inference?

c. What will happen to the lake if the water continues to pour in?

Experience 4: OBSERVING STREAM DEPOSITION

OBJECTIVE

Observe and describe a delta deposit and alluvial fan.

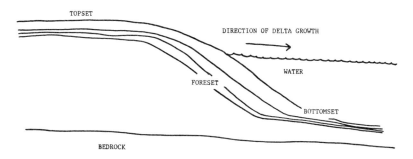

Diagram 18

PROCEDURE

a. Place dirt in one end of a transparent box and elevate that end of the box 2-3 inches. Place some water in the empty end of the container and squirt additional water on the dirt to form a stream. Observe what happens to the dirt moved by the water as it reaches the lake, and record your observations with drawings. (See previous experience.)

b. Repeat the process of forming the delta but slowly add black sand at the source of the stream. After the black sand is deposited, add white sand to the stream. Continue alternating these until a good delta is formed. This produces layers which are easily visible from the bottom or where the delta may touch the edges of the container.

c. Some streams deposit material on land rather than in a lake. Such a deposit on land is called an alluvial fan. To simulate the formation of an alluvial fan, repeat the procedure from a and b using another container, only do not allow a lake to form. This can be done by siphoning out the water as it collects, removing the water with a spoon or placing a hole at the lake end of the container. Add the black and white sand as was done before. Describe the alluvial fan. Describe any likenesses or differences you can observe.

d. Carefully dig into the delta and fan. Describe any likenesses or differences you can observe.

e. Where are the largest particles found in the delta? In the fan?

f. Is there any similarity between the slant or slope of the layers in the delta and fan? If so, explain.

VOCABULARY

alluvial fan

MATERIALS

transparent plastic boxes (or glass baking dish with large base)

plastic squeeze bottle for water

dirt mixture

white sand (available at hardware store, e.g. Sakrete All Purpose Sand)

black sand (mix white sand with black India ink and allow to dry)

pail

spoon (or plastic or rubber tubing)

PROCEDURAL INFORMATION

An important difference between the delta and the fan is that the stream deposits the fan and later runs across what is previously deposited. The stream thus erodes its own deposits on the fan, thereby effectively lengthening its course. This is not the case with the delta. As in the case of the Mississippi Delta, entire towns can be on the river's deposits.

In some deltas three sets of layered beds can be distinguished. These are the foreset, bottomset and topset beds. The foreset beds are made of coarser material dropped quickly, in rather thick layers, on the outer slope of the delta. Each bed is inclined or sloped at the same angle. The bottomset beds are laid down beyond the foreset beds and usually consist of three horizontal layers of finer material that settles more slowly. The topset beds, also nearly horizontal, are the stream deposits placed on top of the delta. As time passes, younger topset and foreset beds advance over previously-laid, older bottomset beds.

The very fine materials deposited as mud in the delta can be compared with the materials comprising shale. Shale is basically compacted mud. The mud and other light weight materials are deposited in the delta if the water is relatively calm. The effect of waves or a strong current can be observed by gently rocking the container back and forth. See Diagram 18.

EVALUATION

a. Describe one similarity and one difference in the ways deltas and alluvial fans are formed.

b. Describe one likeness in the structure of a delta and an alluvial fan.

Experience 5: INFER THE STREAM PATTERN ON AN ALLUVIAL FAN

OBJECTIVE

Construct a stream pattern tracing from an alluvial fan, and infer from a stream pattern the landform used to produce the pattern.

PROCEDURE

a. Briefly list the various landforms you see on the topographic map you have been given.

b. If you did not have an alluvial fan on your list, re-examine your map until you find one. Describe the location of the fan in relation to nearby land features.

c. Using tracing paper, trace the stream pattern on the fan. Describe the pattern you obtain.

d. Do you think this stream pattern could develop on any other landform? Explain.

e. Can you think of an explanation for the short streams that seem suddenly to end?

MATERIALS

topographic maps showing well-developed alluvial fans (e.g. Ontario, California or Ennis, Montana)

tracing paper

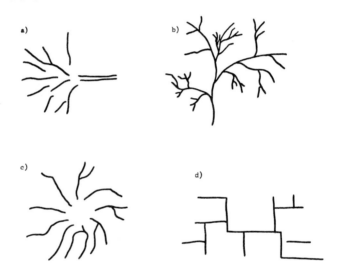

Diagram 19

PROCEDURAL INFORMATION

An alluvial fan is composed of stream-deposited materials which are quite porous. A stream without a strong source of water can often "soak in" and disappear from the surface. This explains the short streams which do not empty into a larger body of water. Also, the intermittent stream symbol —···—··— may be found on alluvial fans. This indicates a stream which is dried up at certain seasons.

This experience may prove difficult for some students and

might best be done as a class activity with additional direction. Another alternative would be to have the students make tracings of a variety of river patterns leading up to the stream on the alluvial fan. The mystery of streams which appear to have no mouth will be intriguing for the students to solve.

EVALUATION

Identify the stream pattern which is associated with the alluvial fan. See Diagram 19.

Experience 6: IDENTIFY STREAM CHANGES IN THE FIELD

OBJECTIVES

a. Measure the size of the particles being transported and those being deposited.

b. Describe the shape of particles found in the stream bed.

c. Identify and name erosional and depositional features found in the stream.

d. Observe the processes of change taking place in the stream and infer the causes of the changes.

PROCEDURE

a. Where the stream's velocity changes look for:

Changes in the size of rocks transported.

Changes in the size of rocks deposited.

Changes in the depth of the channel.

Reasons for changes in velocity.

b. What effect does a second stream, or tributary, have on the first stream?

What occurs at the streams' junction?

Is there a change in the stream's volume? velocity? channel depth?

Do transported or deposited particles vary in the two streams?

c. If the stream contains a bend or meander:

Where does the stream travel slowest when going around a bend? swiftest?

Compare channel depth and sand bars on the inside and outside of the meander.

d. If the stream contains a waterfall:

Why does the falls exist?

Check for layering in the rocks and observe any differences in the structuring of the layers.

Identify unique features in the falls area.

e. Is there a flood plain? What is it made of? How do you think it was formed?

f. Are both sides of the stream wearing away equally where the stream is straight?

MATERIALS

jars for water samples	tape measure
hammer (optional)	rulers
magnifying lenses (optional)	first-aid kit
topographic map of area	camera (optional)

PROCEDURAL INFORMATION

The need to move from the classroom model to the field quickly becomes apparent, once at the site of a stream. The students may be expected to have difficulty relating what they learned from the classroom models to the real situation. It is wise to start exploring the more obvious aspects of the stream first. It may take a prolonged visit, or several shorter ones, before teacher and students feel comfortable with their stream and its characteristics.

It may be desirable to divide the students into teams of 2-3 students each. Each team could be responsible for determining the stream flow and particle size, identifying features indicating erosion and deposition, sampling the water for transported material, and examining the general shape of particles. Or, each team could examine one of these aspects and share its findings with the rest of the class.

A permanent stream would be ideal to visit. However, a small temporary or intermittent stream could be used. Such a stream would be found immediately following a rain and could flow over a field, sidewalk, parking lot, or along a curbside. The same procedure would be used in studying either type of stream, with the determination of which type to use being a decision of the teacher.

EVALUATION

Given a field situation, the student will complete those of the following which are appropriate for the given stream:

a. describe the particles found in the stream bed and compare their sizes with any materials being transported.

b. identify evidence of two effects of erosion or deposition in the stream.

c. state one inference regarding a change taking place in the stream and its cause.

4

How Do Maps Show the Size
and Shape of Landforms?

One of the most relevant types of maps for students' use is the topographic map. Over two-thirds of the United States has been mapped by the U.S. Geologic Survey. These maps vary in detail depending upon the density of the area shown. However, inexpensive topographic maps with a ratio of 1:24,000 are available for many of the more densely populated areas. Streets, houses, churches and schools are indicated as are bodies of water, woodlands, and even wells in the arid regions.

Children are intrigued with topographic maps of their locale showing the familiar landmarks. This interest can serve as the motivation for learning some of the processes of representing land sizes and shapes on paper. These maps may be used in many areas of the curriculum. The influence of topography on people and communities can be observed. The mathematical relationship between the actual land mass and its representation on the map is a realistic application of scales and ratios. Those concerned with the environment will find that woods and swamplands, even hills and valleys have disappeared only to be replaced by new streets, buildings, and entire communities.

Topographic maps can best serve those who understand their intricacies through experience and use. They contain a vast amount of information. Through the actual construction of their own topographic maps, children will become more adept at recognizing and using this information and their teachers will find many uses for topographic maps. Using topographic features or maps of the local area will increase the value and relevancy of these experiences for the children.

Experience 1: DRAWING TO SCALE

OBJECTIVES

    a. Construct a map of a given area.

    b. Construct a map to scale and name the scale.

PROCEDURE

    a. Draw a map of the classroom showing the desks, tables, doorways, windows, and other large pieces of furniture.

    b. Compare your map with those drawn by other students and discuss the reasons for any differences. How could many of these differences have been eliminated?

    c. Construct a map of the classroom on a sheet of graph paper. Decide on a scale for your map and draw each feature to that scale. Indicate the scale.

    d. Compare your map to those drawn by other students and discuss the differences. Is the variation as great as it was in the maps that did not use the scale?

    e. From the maps you have just drawn, tell why drawing a map to scale is important.

VOCABULARY

    map   scale

MATERIALS

    tape measure
    graph paper having divisions of at least ¼ inch

PROCEDURAL INFORMATION

    Being able to show a person how to get to a given location is a skill. It requires that one be a keen observer and be able to communicate his observations to another person. The easiest way to do the latter is with a simple map, indicating distances if not drawn to scale. The child who can develop such a map is well along the path of being able to use maps made by others in his activities.

This experience should be used to introduce the idea of using scale which the students will need in following experiences. The time required for this will vary considerably, depending upon the students' previous experiences with drawing diagrams to scale. The emphasis should be on size and distance relationships on the maps rather than the inclusion of all features in the room or details of these items. A discussion of how a map can be used to communicate when a scale is included might be helpful. The differences between a map of the room and a picture of the room may, also, be included.

EVALUATION

Given a sheet of graph paper and a tape measure, draw a scale map of a portion of the classroom or a small area outside, as assigned by your teacher. Label the map and indicate the scale.

Experience 2: REPRESENTING DISTANCES ACCURATELY

OBJECTIVES

a. Construct a scale map of a given area and name the scale.
b. Measure distances between given locations on maps.

PROCEDURE

a. Working in small groups, draw a scale map which includes the school buildings and immediate surroundings, as designated by the teacher. Use the tape measure to determine distances. What will 1″ on your map be equal to in the field? (Check your scale with your teacher before actually making your map.)

b. Measure the length and width of your school building and

other distances shown on your map as your teacher indicates them. Compare your measurements with those of some of your classmates' maps, using their scale. Do the dimensions agree with yours?

c. Compare your map with those of your classmates to see if they are alike. Your map was made using direct measurement. Are all maps made this way? Explain.

MATERIALS

graph paper   tape measure   ruler
table with a few objects placed on it (for evaluation)

PROCEDURAL INFORMATION

Decide on the area to be mapped in advance. The school building does not have to be included if you prefer to use other areas. Streams, athletic fields, playgrounds, or part of the school neighborhood may be preferable for individual situations. If it is impossible to map outdoors, a model may be set up with building blocks or toy buildings for mapping on a smaller scale. It is important that areas to be mapped are checked in advance to eliminate any difficult or problem areas.

Make sure each group of students is using a reasonable scale early in the activity. Students may determine the length of their pace and use this to measure distances rather than using a tape measure.

A possible additional activity would be to have each student measure and determine distances using the scale between points on a road map.

EVALUATION

Given a table with objects arranged on it, construct a scale map of it and name the scale. State the distances between a few objects on your map. (Teacher identifies 3-4 objects.)

Experience 3: IDENTIFYING CONTOURS ON A DELTA

OBJECTIVE

Construct a topographic map of a delta.

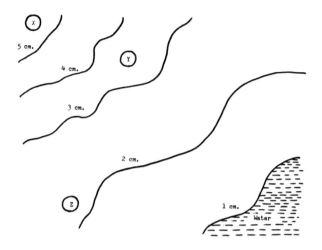

Diagram 20

PROCEDURE

a. Make equally spaced markings about 1 cm. apart on the outside of the plastic box with the marking pen. The marks should be from the bottom to the top of the box so you can put 1 cm., 2 cm. . . . of water into the box. Fill one end of the box with dirt.

b. Set up the materials (as described in chapter 3, Experience 4) and let a delta form. Do you see anything happening which you did not see before? Let the water flow until the soil is just covered by water.

c. After the delta is formed, very carefully remove the lake which has formed, by siphoning or spooning out the water.

d. Draw a map of the landform you have just made in the box, including the delta. You should be concerned about showing hills, valleys, changes in height, and the general topography.

e. Without disturbing the dirt, pour enough water into the tray to bring the water level up to the first mark on the side of the box (1 cm.).

Place the toothpicks into the dirt at the water's edge so that a fence is made which separates the land and water. The toothpicks should be about ½ cm. apart.

Are the toothpicks you've just placed all at the same distance above the tray bottom?

Explain.

f. Put the top on the plastic box and place the clear plastic sheet on the top. Using the grease pencil, trace the outline of the toothpick fence on the plastic sheet.

g. When you have completed the outline, remove the cover and again add water to the box, this time making the water 2 cm. deep. Place toothpicks in the dirt at the water's edge as you did before and trace the outline of this fence on the plastic sheet.

h. Repeat the process used in e, f, and g, until all of the land is covered with water.

i. How much vertical distance is there between each row of fence? Label your map, showing which line is 1 cm. above the bottom, 2 cm., and so on. These lines are called contour lines and the elevation between them is the contour interval. In this case, each line represents an increase in elevation of 1 cm. so the contour interval is 1 cm.

j. Does this map show the delta to be higher on one end than on the other? Explain.

k. Does your land have a hill or valley? How is the hill or valley shown on your map?

l. How many centimeters above the bottom of the box is the highest part of the land? Could you answer this question if you had only the first map prepared in d? Explain.

VOCABULARY

topography   contour lines
delta   contour interval

MATERIALS

plastic (shoe or sweater) box with cover
plastic squeeze bottle
plastic or rubber tubing or spoon
transparent acetate sheet 8½" x 11", 1 per group
masking tape (to hold acetate in place; strip may be placed
vertically on outside of box to make the 1 cm. marks)

dirt sample   toothpicks
water source   drawing paper

PROCEDURAL INFORMATION

Starting with a landform (delta), the student marks different
elevations above the bottom by progressively filling the tray
with water and marking the new beach with toothpicks. The
vertical distance between each toothpick row or line of the
plastic sheet is the same. This vertical difference is called the
contour interval. The lines themselves are called contour lines.

EVALUATION

a. The students may be given the materials to make a delta or
alluvial fan and be asked to make a topographic map showing
labeled contour lines with 1 cm. contour intervals.

b. Provide the students with a map similar to Diagram 20 and
ask such questions as:

What point is higher, X, Y, or Z? Explain.

If the water would rise 3 cm., which land areas would be
submerged? (1) none (2) X only (3) Z and Y (4) Z only

Which is the point of highest elevation on the map? (1) Z
(2) Y (3) X (4) water surface

What is the contour interval of this map? (1) 5 cm. (2) 4
cm. (3) 2 cm. (4) 1 cm.

**Experience 4: IDENTIFYING DIRECTION, SCALE
AND MAN-MADE OBJECTS**

OBJECTIVE

Construct a topographic map showing contour lines, map scale, map direction and symbols, and demonstrate its use.

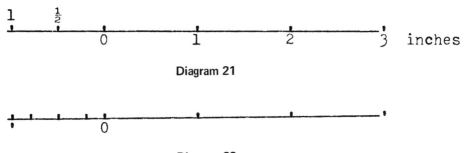

Diagram 21

Diagram 22

Feature	Symbol	Location	Elevation
Road		Running N-S 2" from map's eastern edge	
House		1 house on top of hill	
Church		1" from N and 1" from W boundry	
School		2" S from church	
Cemetery		¼" E from church	

Diagram 23

PROCEDURE

a. Make a hill or mound of dampened dirt in the center of the plastic box. The hill should be much steeper on one side than the others. Pack the dirt so it won't easily wash away when water is added.

b. Using a compass, place the plastic box so that one end faces north, the other south, and the sides are lined up with east and west. Sketch the box and label the four compass directions on your paper.

c. Pour enough water into the box to make the water level
1 cm. above the bottom of the box. Cut the plastic sheet to the
same size as the cover of the box and attach it to the top of the
cover with masking tape. Place the top on the box and trace the
water's edge on the sheet of plastic.

d. Repeat step c, adding 1 cm. of water each time, until the
hill is just covered with water and each new water edge is traced
and labeled.

e. Label the four compass directions as well as the level of
each contour line on the plastic sheet.

f. Compare the contour lines on the steep and gentle sides of
the hill. Why are the lines closer together in some areas?

g. Look at the line labeled 2 cm. What can you say about
every point on that line?

h. How is the hill shown on the map?

i. Your map is the same size as the land it is representing.
That means 1″ on the map is the same as 1″ on the land. This
can be written as 1:1 which would mean 1″ on the map = 1″ on
land. Place this scale and a short representation of the scale on a
line such as in Diagram 21 on your map.

If a scale of 1:1 means 1″ on the map = 1″ on land, what
would the ratio 1:2 mean?

j. What would you have to do in order to make your map
have a scale of 1:2? Would a 1:2 scale represent more or less
land than a 1:1 scale? How much more or less land?

k. Label the line in Diagram 22 the way it would appear if it
were on a map with a 1:2 scale.

l. The land is seldom bare of man-made features. Buildings,
roads, and cities are present and a map that accurately shows
the land should also show these features. Using the symbols
given in Diagram 23, place the features on your map where they
should be.

m. After you have placed the symbols on your map,
determine the approximate elevation of each feature. Remember that all points on a contour line have the same elevation. If
a feature is not on a line, use the closest line to the feature and
record the elevation on the chart.

VOCABULARY

symbol   scale   contour line

MATERIALS

plastic box   ruler
dirt   compass
water   masking tape
transparent acetate sheets, 8½ x 11 inches
felt tip pens or wax pencils
scissors

PROCEDURAL INFORMATION

The student should be able to recognize that his map is the same size as the land it represents. You might discuss the possibilities of making a map of the state using a 1:1 ratio and the problems related to such a map. For this reason, most maps are physically much smaller than the area they represent. It may be helpful to examine the scales used on road or topographic maps. Most topographic maps use 1:62,500 or 1:24,000 scales.

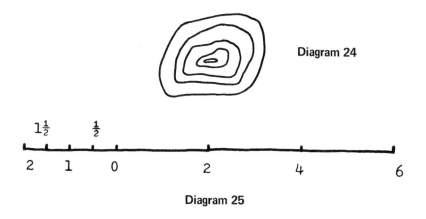

Diagram 24

Diagram 25

A contour line connects points of equal elevation. All points on a contour line therefore, have the same elevation. A hill would be shown by progressively smaller concentric contour lines, one inside the other with increasing elevations. The closer together the contours the greater the change in elevation. On the steep side of a hill the lines would be closer together than on the rest of the hill. See Diagram 24.

If the scale on a map of 1:1 means 1 map inch = 1 land inch, then a 1:2 scale means 1 map inch = 2 land inches. The map, therefore is one-half the physical size of the land it represents. For the students to make a map with a 1:1 scale into one having a 1:2 scale, they would reduce their entire map by one-half. This would enable them to use a piece of paper one-half as large as that previously required. See Diagram 25.

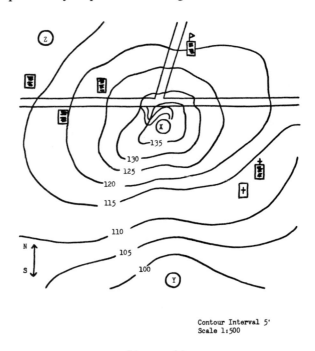

Contour Interval 5'
Scale 1:500

Diagram 26

EVALUATION

    a. See Diagram 26. Is there a hill on the map? Explain.

    b. Which point, X, Y or Z, has the lowest elevation?

    c. How long is the road running N-S?

    d. There is no road going to: (1) the school (2) the church (3) the hill top?

e. What is the approximate elevation of the school?

f. If you were asked to build a road to the church at the cost of $10/inch, how much would the least expensive road cost?

Experience 5: IDENTIFYING WHY CONTOURS BEND

OBJECTIVE

Demonstrate why contour lines seem to bend or point when crossing a stream.

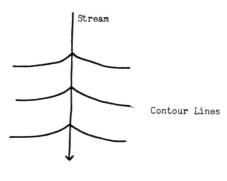

Diagram 27

PROCEDURE

a. Examine a topographic map until you find a stream. What happens to a contour line when it crosses the stream?

b. Draw a picture that shows the stream, the contour lines crossing the stream, and the direction in which the stream is flowing.

c. State a rule that seems to describe how contours show which direction is upstream or the direction from which the stream is flowing. Observe other streams on the map to see if your rule applies to them, if not alter it so it does.

d. Can you think of an explanation for why the contour lines point upstream? Explain.

e. Place some dirt in the plastic box and either dig out a river

valley or let one form naturally. Drain the excess water from
the box. Carefully add water, increasing the depth 1 cm. at a
time while you map the elevation of the land as in Exercise 3,
using the box cover and transparent acetate. Draw the contours
with a felt tip pen (or wax pencil) until all of the land is covered
with water.

f. Describe what you observed, keeping in mind that you
should be able to explain why contour lines point upstream.

MATERIALS

plastic boxes   water
dirt   spoon
any topographic map showing streams
transparent acetate sheets
felt tip pens or wax pencils

PROCEDURAL INFORMATION

Streams are usually found in the valleys or gullies they have
carved for themselves. This process has been previously investi-
gated. By definition, a contour line connects points with the
same elevation. If a contour line were able to approach a stream
and go straight across it, the line would dip into the valley of
the stream and, therefore, be at a different elevation. For a
contour line to continue showing the same elevation at this
intersection, the contour must turn upstream where it is higher.
Therefore, contours point upstream. The stream movement can
thus be determined by simply looking at the contour lines
where they cross the stream. This is helpful, especially when
only a portion of a stream or river appears on a map.

The students' diagrams for step b should resemble Diagram
27.

EVALUATION

Given the topographic map in Diagram 28, answer the
following questions:

a. If you wished to leave Lake Luise in a canoe and go
downstream to the northwest, which stream would you take?

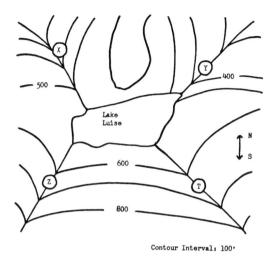

Contour Interval: 100'

Diagram 28

(Answers: (1) X (2) Y (3) Z (4) T)

   b. Which streams enter Lake Luise?

   c. Which streams leave Lake Luise?

   d. Stream T flows in which compass direction? (Answers: (a) X (b) Z and T (c) X and Y (d) NW)

Experience 6: CONSTRUCTING CONTOURS ON A SMALL AREA

OBJECTIVE

   Construct a topographic map of the schoolyard or a small hilly area.

PROCEDURE

   a. The teacher will explain what is to be done in this exercise and assign various responsibilities to students.

   b. You will be mapping the schoolyard or a small hilly area. Buildings, trees, and contour lines will be measured or determined and placed on your map in their relative position with respect to each other.

c. Don't forget to adopt an appropriate scale and orient objects in their proper position with the compass.

VOCABULARY

level   surveyor's level

MATERIALS

carpenter's level (If unavailable, take a smooth-surfaced, clear bottle with straight sides, almost fill with water and cork tightly. When it is tilted on its side, the bubble will act much as it would in a level. Nail a piece of wood about the same length as the bottle to the top of the long board to form a T. After the bottle is taped into the position shown, the top edge of the bottle should be at the students' eye level when steadied perpendicular to the ground. See Diagram 29. The students may have to bend down to use this level or Jacob's staff as it is often called.)

large piece of cardboard (to fasten newsprint on while in field)

to make level: bottle, 1 piece 1″ x 2″ or 2″ x 4″ board about 4′ long, 1 piece of 1″ x 2″ board about 1′ long

masking tape	markers (colored cloth on nails)
tape measure	hammer
compass	pencils

large sheets of newsprint

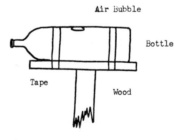

Diagram 29

PROCEDURAL INFORMATION

The level serves to keep the stand level with relation to the earth. When held level, the student can sight along the top of the level and the land he views will be at the same elevation as the top of the level. See Diagram 30. The height of the sighting stand can be varied allowing for changes in the contour interval. If the schoolyard is not too hilly, a small interval, perhaps 2', would serve. If the stand is made 2' tall, then the land sighted will be 2' higher than the ground the level stands on. A 2' level may be more difficult for the students to use than a 4-5' one depending on the size of the students.

To begin, the lowest place in the yard is located by estimation. The level is stood upright in the low spot and supported by a student. A second student can assist by making sure the instrument is held level. These students mark the original spot where they placed the level with a marker (a piece of red cloth on a nail). As the first student sights along the level, a third student places a marker on the sighted spot so that it can be easily seen. The students sighting rotate the level and indicate other areas of similar elevation. The student marking the contour now moves to these sights on the contour and marks them, continuing until the entire contour is marked.

When the first contour is laid out, the sighting stand and sighters move to a spot on the first contour and proceed to lay out the second contour, then the third, until all of the area has been mapped. As contours are completed, other students can be assigned the task of recording them in their proper place on the map. The tape measure can be used to determine the distances from various points on the contour to prominent objects such as trees, buildings, or large rocks. The compass will give the direction from the object to the contour. Students may thus find that the first red marker placed down was 17' due south of the maple tree. By placing the tree and other objects on the map in their respective locations, the contour lines can be drawn in.

This project will probably involve the entire class as opposed to 2-3 student teams, at least for the first effort. Six to eight

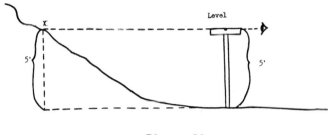

Diagram 30

students could be placing markers, while others are measuring distances, and still others are actually drawing up the map.

If the schoolyard is not sufficiently hilly to make the experience interesting, a short walk or bus ride might reveal a park or hill which could be used.

Comparisons with this experience and those using the model, where the land was slowly submerged to determine the contour, would be helpful prior to taking to the field. It is important for the students to recognize what they are going to do before going into the field. The organization of the materials and teams for different tasks such as marking or measuring is a good experience for them.

ADVANCED PREPARATIONS

It might be wise to determine an appropriate contour interval before the level stand is constructed so it will be at a convenient height. This might be constructed by members of the class or students in a shop class. If comparisons are to be made with U.S. Geologic Survey maps, it might help to select an interval which would lend itself to easy comparison. Discussions on the differences in detail and changes would be aided by such forethought.

EVALUATION

a. Describe how a topographic map is made.

b. Complete the topographic map (Diagram 31) by drawing in the contour lines. The numbers represent determined elevations.

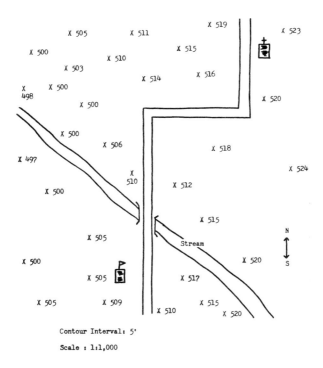

Diagram 31

Experience 7: CONSTRUCTING A TOPOGRAPHIC PROFILE

OBJECTIVE

Construct a topographic profile.

BACKGROUND

You have seen how a land area can be mapped on a piece of paper and still show the topographic features of the land. Can the reverse be done? In other words, can you start with a map and reconstruct the topographic hills and valleys? To do so is to make a profile or skyline view of the topographic map.

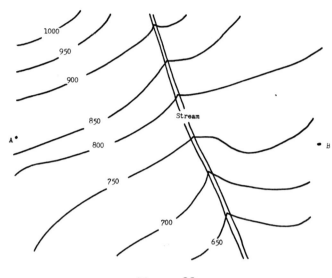

Diagram 32

PROCEDURE

a. Find letters A and B on your topographic map (Diagram 32) and draw a line from A to B.

b. Check the contour lines along line AB and determine the highest and lowest elevations line AB crosses:

Highest:_____

Lowest:_____

c. Label the edge of a piece of graph paper with elevation numbers so that the highest and lowest values of the map may be included. The numbering should progress in consistent intervals, e.g. 625, 650, 675, 700 or 600, 650, 700, 750.

d. To construct a profile along line AB, place one edge of your graph paper on line AB and mark it on the location of A and B.

e. Starting at A, move along line AB until you come to a contour line. Then, using a ruler, make a vertical line from the contour to the previously labeled line on the graph paper which has the same elevation as the contour and place an X where the 2 lines meet. Continue along line AB until this has been done for all of the contours.

f. When you have worked all the way from A to B, connect

the X's you have made on the graph paper and you will have a side view or skyline view of the land shown on the map along line AB. This side view is called a topographic profile. Remember that all points between 750-800-750 are higher than 750, therefore hump; all points between 750-700-750 are less than 750 and therefore dip.

g. When you finish, draw another line, perhaps in a different direction across the map and make a second profile.

VOCABULARY

profile

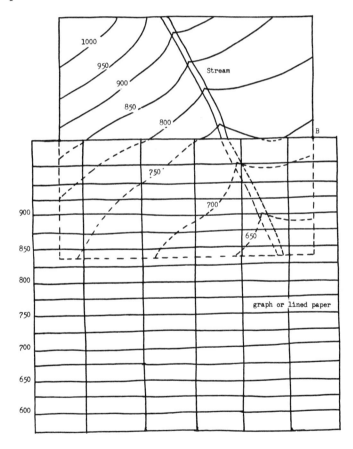

Diagram 33

MATERIALS

topographic map similar to Diagram 32 (real or fictional)
ruler
graph paper (regular lined paper may be used)
topographic map (for evaluation)

PROCEDURAL INFORMATION

The series of diagrams (Diagrams 33, 34 and 35) shows the sequence to be followed by the students in steps c, d, e and f. The interval selected for step c will depend upon the size of the graph paper used.

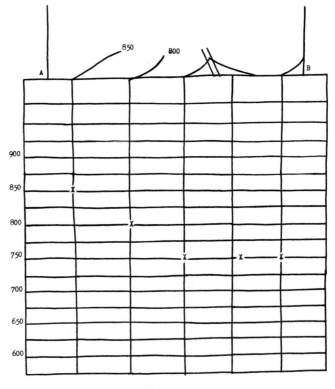

Diagram 34

Students have a tendency to draw a straight line when two or more side-by-side points on their graph paper have the same elevation. For example, the last 3 contours to be crossed when approaching B on line AB all have an elevation of 750'. The students should recall from previous experiences that the land would not be flat since a stream is present. A discussion should reveal that it is not known precisely what the elevation would be in cases such as this but that one can infer what it would be based on past experiences and knowledge. As one becomes more familiar with different landforms, such inferences should be more reliable.

Profiles can be made in any direction and students may find it interesting to compare a variety of the profiles which they make.

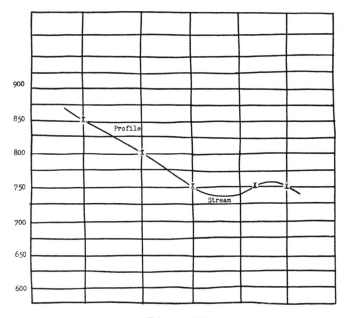

Diagram 35

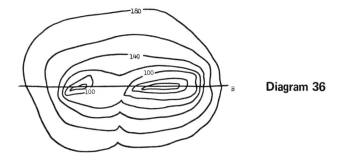

Diagram 36

EVALUATION

Given a contour map Diagram 36, graph paper and a ruler, construct a topographic profile along line AB.

Experience 8: DEMONSTRATE THE USE OF A TOPOGRAPHIC MAP

OBJECTIVE

a. Construct a stream pattern tracing from a topographic map.

b. Infer the direction of stream flow and the possible source of the water.

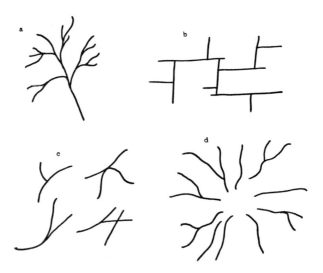

Diagram 37

PROCEDURE

a. Answer the following questions from the topographic map you have been given.

1. What is the name of the topographic map?

2. When was it made?
3. Where is the land located that the map represents?
4. What is the scale used on this map?
5. What is the contour interval?
6. What is the outstanding feature near the map's center?
7. What is the shape of this feature?
8. Are there any streams on the mountain?
9. What appears to be the source of some of these streams?

b. Place your tracing paper over the mountain and trace the stream beds on the mountain. Do the streams run up or down the mountain?

c. Describe the pattern made by the streams. Why do you think the streams follow this pattern?

VOCABULARY

drainage pattern   drainage basin
              volcanic mountain

MATERIALS

topographic map of Mt. Shasta, California; available U.S. Geologic Survey, Washington, D.C.
tracing paper

PROCEDURAL INFORMATION

The first few questions are to review some of the information included on the map. Both the map name and date of production are found in the bottom right corner of USGS maps. The map scale and contour interval are found near the bottom center of the map.

The most prominent feature on this particular map, of course, is the volcanic mountain. The mountain is not presently active but it is volcanic in origin. The shape resembles that of an inverted cone. The mountain glaciers supply sufficient water to maintain some of the streams. It may be noted that the north side of the mountain has more glacial development than does the south side. This is the result of the more direct sunlight on the south side while the north side is the shadow side.

Reasoning should suggest that the streams run down the mountain and this is reinforced by the pointing upstream of the contour lines. The cone shape causes the stream pattern to radiate like spokes of a wheel from the central peak. This type of stream pattern is know as radial. Only mountains with conical shapes have a radial stream pattern, therefore, this pattern serves as a clue to the presence of mountains on a map.

EVALUATION

a. Given a different volcanic mountain, e.g. Mt. Hood, Oregon, the student is able to make a tracing of the stream pattern, infer stream direction and possible source of water.

b. Diagram 37 shows four stream pattern tracings. Which was made from a cone shaped mountain?

5

Investigation of the Relationship Between Light and Color

Variations of color and light add interest to life, especially to young people, as evidenced in their preference for bright colors and colored lights. This interest can carry over into the classroom through observing the effects of light on color and the relationship between the two. The application of these relationships to nature helps to explain the rainbow, the colorful sunset, and even the ominous darkness before a storm.

Children find the activities which permit them time to try things on their own interesting. This is especially true if they are given ample time to try out a prism or the effects of colored light on other colored objects. Their inquiring into "what would happen if" types of situations will produce answers, raise questions and be fun.

Experience 1: DEMONSTRATE THE FORMATION OF A SPECTRUM

OBJECTIVE

Demonstrate how a spectrum is made and identify and name the colors in a spectrum.

PROCEDURE

a. Place a prism on a flat surface in front of a light source. Cut a long, narrow (10 cm. x 3 mm.) slit in a large index card and hold it between the prism and the light source. Some of the light should pass through the slit to the prism and through it onto a nearby wall, screen or piece of paper. See Diagram 38.

b. Record the variety of colors and their position in the spectrum.

c. Does moving the prism change the colors or their position in the spectrum?

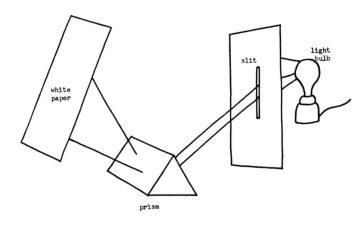

Diagram 38

ALTERNATE PROCEDURE

a. Fill an aquarium with water and place it near a filmstrip projector as shown in the diagram. Cut a long, narrow (10 cm. x

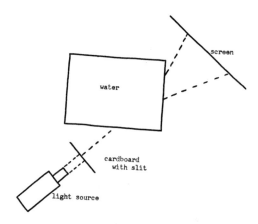

Diagram 39

3 mm.) slit in a large index card and hold it between the aquarium and the projector. Light from the projector should pass through the slit to the aquarium and through it to a wall, screen or piece of paper. See Diagram 39.

b. The aquarium acts as a prism does, separating the light into the colors of the spectrum. Record the colors in the spectrum in the sequence you see.

c. Does moving the projector change the colors or their position in the spectrum?

VOCABULARY

prism   spectrum

MATERIALS

large index card   scissors
metric ruler   light source
     triangular prisms (if not available, use one of
         the alternatives shown)

PROCEDURAL INFORMATION

This experience is basic to the study of light and color. Sunlight is more desirable but the projector may be used if sunlight is not available. In either case, the light should be confined to a narrow beam of light by using the cardboard. A discussion of the spectrum produced should result in identifying the visible colors of the spectrum as red, orange, yellow, green, blue, indigo and violet. Indigo may be unfamiliar to the students.

Two additional alternatives are shown in case a prism is not available. Prisms are inexpensive and require less fuss than the alternatives. After using the prism, some students may want to try one of the alternative methods for comparison. See Diagrams 40 and 41.

The water acts as a prism in that it affects the passage of light as glass does. However, the light must enter the water at a small angle in order to have the light exit from the water to form a spectrum. This effect is directly related to rainbows in the atmosphere, near waterfalls and around water sprayed from a hose.

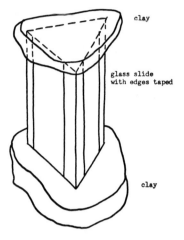

Diagram 40

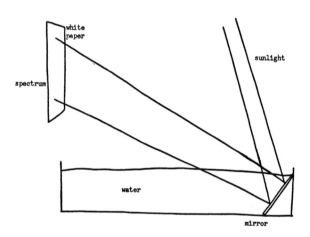

Diagram 41

The students should be given ample time to "play" with the materials and explore the spectrum produced. It may be pointed out that the light rays in white light are bent differently as they pass through the prism. The colors are all present in white light but the prism acts as a separator by alternating the angle at which the rays leave the prism. The students may conclude that there is a predictable order in the degree of bending which results in the color arrangement of the spectrum, i.e. red light is bent less than violet light.

EVALUATION

a. Given a prism, a cardboard with a slit in it, a light source and a piece of white paper, demonstrate how a spectrum is made.

b. List the colors of the spectrum in order as they appear in the spectrum.

Experience 2: PRODUCING WHITE LIGHT FROM A SPECTRUM

OBJECTIVE

Demonstrate that a blending of all colors in the spectrum will produce white light.

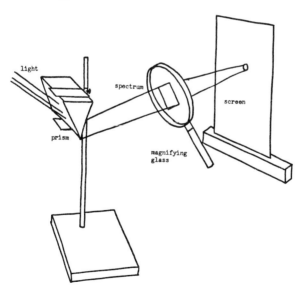

Diagram 42

PROCEDURE

a. Using a glass prism, form a spectrum on a screen using either sunlight or light from a projector. For best results, fasten the prism on a ringstand so it will remain in position. See Diagram 42.

b. List the colors obtained from the white light.

c. Place the magnifying glass near the screen where the spectrum is. Slowly move the magnifying glass towards the prism while watching the light on the screen. Observe any color changes in the spot of light on the screen.

d. Is it possible to mix the colors of the spectrum to produce white light? Compare your results with others in the class.

VOCABULARY

lens   refraction

MATERIALS

triangular glass prism
magnifying glass
sun's rays or light from a slide projector
ringstand and clamp
screen (sheet of white poster paper)

PROCEDURAL INFORMATION

A well-defined spectrum should be obtained before starting step c. The students should be cautioned to be sure that the magnifying glass traces the path of the light rays producing the spectrum as they move it towards the prism. They should be able to focus the light beam of the spectrum into a small white circle on the screen.

EVALUATION

The student will describe how to obtain white light from a spectrum using a white light source.

Experience 3: CONSTRUCTING A COLOR WHEEL THAT PRODUCES
WHITE LIGHT

OBJECTIVE

Construct a primary color wheel which, while spinning, appears to be nearly white, name the primary colors used, and distinguish between white and black color in terms of light.

PROCEDURE

a. To make the color wheel, draw a circle with a 7 cm. diameter on each of the pieces of red, blue, and green construction paper. Cut the discs out, draw a radius on each disc, and cut the length of each radius.

b. Slip the three discs together by interlocking them along the cut radius. Rotate them until 1/3 of each color is showing. Cut another disc the same size from the Manila paper and tape the color wheel to it.

c. Take the bolt and carefully make a hole in the center of the color wheel the same size as the diameter of the bolt. Slip a washer on the bolt, then the color wheel, another washer and then place the nut on the bolt and tighten it so the color wheel and washers are held firmly in position.

d. Place the bolt in the drill and spin the disc. Observe the color of the disc. What color do you see?

e. Is there any difference in the color when the disc is spun at different speeds? Is there any difference in the color if you spin the disc in sunlight or in artificial light?

f. Rotate the disc so different amounts of the colors are visible. What difference does this make when the disc is spun?

g. When spinning a color wheel made of the proper shades of red, green and blue, you should see white or close to it. Different shades of these colors will make a difference in your results. You may want to try using other colors. What do you think would happen if you made a wheel of the colors in the spectrum?

h. What would you do to make the original rotating disc appear black? Try it.

VOCABULARY

light   white light
drill   black

MATERIALS

scissors   drawing compass
metric ruler   Manila paper
hand drill   tape
nut and bolt (threaded all the way to the head)
2 washers (about 1-2 cm. in diameter, must fit bolt)
construction paper (red, blue, green and others)

PROCEDURAL INFORMATION

The selection of the shades of the three colors will influence the degree of whiteness of the disc as will the speed at which the wheel is rotated. Some students may prefer to use a piece of heavy white paper and color or paint in the three sections of color. This would permit them to alter the shades.

Encourage the students to try several color combinations and to record and compare their results. To obtain black from their color wheel, they would have to use black paper or take their color wheel into a dark room. Since, in terms of light, black is the result of the absence of light, any color wheel would be black in the dark room.

This series of activities generally brings up a discussion of the primary colors in art class. These are not the same as the primary colors of light. In art, the students know they cannot mix red, blue and green paint to get white. Involving the art teacher in this discussion will add to the students' interest.

If an electric drill is substituted for a hand drill, extra caution should be used.

EVALUATION

Given a selection of several colors (cloth, paper, paint, or crayons), the student will select the primary colors and construct a color wheel which appears nearly white when rotated.

Experience 4: DISTINGUISH BETWEEN WHITE AND
MONOCHROMATIC SPECTRA

OBJECTIVE

Observe and distinguish between the spectra produced by white light and monochromatic light passing through a prism.

PROCEDURE

a. Cover a section of a window with opaque paper having a narrow slit in it to allow a small beam of sunlight through it. Position a prism in the beam of sunlight so that it forms a spectrum on a screen or white paper. Record the colors in the spectrum.

b. Place a piece of red cellophane over the slit and record the colors of the spectrum produced. It may be necessary to use 2-4 thicknesses of the cellophane. Repeat this procedure using blue and then green cellophane. Compare the spectrum produced in each case. Does there appear to be any pattern between the color of light and the spectrum produced?

c. Try placing the different colors of cellophane between the prism and the screen. Does this have any observable effect on the spectrum formed by the white light?

VOCABULARY

monochromatic light

MATERIALS

window on sunny side of building
prism
heavy paper with long, narrow slit
red, blue and green cellophane
tape

PROCEDURAL INFORMATION

Since the cellophane is not a perfect filter, there will be a range of wave lengths which will pass through it, i.e. red cellophane may give a red to orange spectrum. It is possible to purchase commercial filters but bright-colored cellophane is adequate for this experience.

Some of the students may wish to pursue this experience further using two or more colors of cellophane at a time and comparing the results. If they were to mix the proper combination of red, green and blue filters, they would be able to produce white light. Such success with cellophane would be highly unlikely.

EVALUATION

Shown each of several different spectra, distinguish between the white light spectrum, the red monochromatic spectrum, the blue monochromatic spectrum, and the green monochromatic spectrum.

Experience 5: IDENTIFYING HOW WHITE AND BLACK OBJECTS RESPOND TO COLORED LIGHT

OBJECTIVE

Distinguish between white and black in terms of light and construct a chart comparing their response to different colors of light.

color of cellophane	black construction paper	white construction paper
red		

Diagram 43

PROCEDURE

a. Fasten a piece of white and a piece of black construction paper next to each other on a wall in a room which can be darkened.

b. Darken the room. Project the light from a projector on both pieces of paper and observe the colors.

c. Place a piece of red cellophane in front of the projector lens so that the red light is projected on both sheets of paper. Observe, identify, and name the apparent color of the construction papers in the red light. Record the results on the chart in Diagram 43. Using several colors of cellophane, complete the chart.

MATERIALS

black construction paper, 22 cm. x 30 cm.

white construction paper

cellophane, several colors such as red, orange, blue, green, violet

charts for observations

white sand

black sand (dye with black India ink)

PROCEDURAL INFORMATION

If the cellophane colors are not intense, fold the cellophane and use several layers. It is possible to photograph sheets of colored construction paper with slide film to obtain slides which could be used, rather than the cellophane. Don't worry about focusing since the slide will not contain writing or images that must be in focus. This permits the use of any camera and sunlight as the light source for the slides.

Using the information obtained from this investigation, the student should be able to infer that a white body reflects all colors in the same proportion, i.e. it appears white in white light and blue in blue light. A black body reflects no color, it absorbs each color shown on it and appears black, regardless of the color of light directed on it.

EVALUATION

Given a pile of black sand and a pile of white sand in a darkened room, the student will be able to identify the black pile and the white pile after observing both in no more than three different colors of light.

Experience 6: INFER THE SOURCE OF COLOR IN OBJECTS

OBJECTIVE

Observe the effect of different colors of light on objects of different color, and predict the color changes that will occur and infer the source of color in objects.

PROCEDURE

a. Fasten sheets of colored construction paper in a long row on the classroom wall and number each sheet. Include one sheet of white construction paper in the row.

color of light	Observed Color of Number of Construction Paper						
	1	2	3	4	5	6	7
white							
red							
orange							
yellow							
green							
blue							
violet							

Diagram 44

b. Darken the room and direct the light from a projector on the row of paper. Record the color of the paper of each numbered sheet on the chart in Diagram 44.

c. Place red-colored cellophane in front of the projector lens and shine the light on the construction paper display. Record your observations.

d. Repeat this process using different colors of cellophane and recording your observations.

e. Examine your recorded observations. Were there times when the color appeared to be the same as it was in white light? What can you infer about the source of the color of an object in relationship to the color of light?

MATERIALS

projector
cellophane or slides, red, orange, yellow, green, blue, violet
construction paper, white, red, orange, yellow, green, blue, violet
chart
room which can be darkened

Food	Color				
---	white	yellow	blue	green	red
bananas					
tomato					
cucumber					

Diagram 45

PROCEDURAL INFORMATION

The chart for the students may be adapted to conform with the colors of cellophane or construction paper available. Also, it may be desirable to have double columns under each color with the student's prediction and the second for the actual observed color. It should be noted that most objects that appear to be of one color are really combinations of colors, one of which is predominant. You may expect variations in the results if you are using inexpensive cellophane and a poorly-darkened room.

Additional student interest can be generated by observing different students and colored clothing under different colors of light. They can also relate this to the effects of colored lights used in plays or at dances.

EVALUATION

Given the chart in Diagram 45, predict the color of the food item under each of the different colors of light.

Experience 7: OBSERVING VARIATIONS IN SOIL COLOR AND COMPOSITION

OBJECTIVE

Observe and identify differences in soil colors of both surface and subsurface samples and compare the composition of the different color soils.

PROCEDURE

a. Collect several samples of soils of different colors. Get at least a cupful of each sample and label the soils by their source.

b. Place a spoonful of each sample on a piece of newspaper. Sprinkle a few drops of water on each sample and observe any color changes. Describe the effect which water has on the samples.

c. Add a few drops of water to each of the samples. Rub each sample between your fingers and record your observations. Do soils that are about the same color feel the same? In the same way, compare the wet sample with some of the same sample which is still dry.

d. Using equal volumes of each dry sample, examine the size of the particles by sorting each sample with the sieve set. Compare the varying amounts of gravel and other large particles, the coarse sand, fine sand, and very fine clay or silt which make up each sample. Construct a chart and record your observations. Is there any relationship between the color and the amount of sieved materials which a sample contains?

e. Describe at least one situation, when a scientist might need to identify the composition of a soil, when its color is the only information he has.

MATERIALS

containers for collecting samples
sieve set, with 4-5 sizes of screening
trowel or shovel
water
newspapers
photos of earth from space (optional)

PROCEDURAL INFORMATION

Soils vary considerably in colors, ranging from the white beach sands to the black swamp muck. It is possible for geologists to make quite accurate predictions of soil compositions from photographs taken from space. If fortunate, they have more clues than just soil color available to them. It is also possible to determine rainfall patterns on earth from such photographs as a result of color differences in wet and dry soils. If colored photographs of the earth or moon are available, the students may wish to compare the colors of soils with known information about the soils in areas shown.

A field trip to an excavation site for a road or building will also reveal color differences in soil. The color variations in the situations would be helpful in identifying the layers of topsoil and subsoil covering the bedrock.

A further extension of this investigation would be to have the students write to students in other states requesting samples of topsoil. A heavy-duty plastic bag, inside a self-addressed stamped envelope, could accompany the letter of request sent in care of the Elementary School principal of the community selected. A discussion of the variations in soil colors may lead some students into some very interesting geology studies.

EVALUATION

From eight soil samples of varying colors, the student will select two samples and make a comparative statement about the composition and color of each, and identify at least one of them by source (surface or subsurface).

Experience 8: IDENTIFYING THE EFFECT OF TEMPERATURE AND COLOR

OBJECTIVE

Observe the color of wire in a flame and infer the relationship between changes in the color and temperature.

PROCEDURE

a. Place one end of the piece of wire firmly into a cork. Light the candle and holding the cork place the free end of the wire into the top of the flame. Observe the wire and all color changes in the sequence in which they occur.

b. Remove the wire from the flame and record any color changes you observe as the wire cools.

c. Repeat steps a and b using a burner in place of the candle. Again, record your observations as the wire is heated and cooled.

d. Are there any similarities in the color changes? Explain.

e. Do you think the color of the wire will change more if it is left in the flame longer? Try it and see. Record your results.

f. What do you infer caused the wire to change color? Based on your observations and inferences, sequence the color changes from coolest to hottest.

MATERIALS

candles   matches
propane burners   corks
pieces of previously heated copper wire
asbestos pads

PROCEDURAL INFORMATION

Caution should be stressed in handling the wires and working with the open flames to prevent fires or burns. The corks act as insulators as the wires are placed in the flame. When the wire is removed from the flame, it should be placed on an asbestos pad to avoid damage to desk tops.

The burner will produce the most changes of color ranging from red to red-orange to orange to yellow-orange. Library research by the students will reveal the colors produced by much higher temperatures.

EVALUATION

When astronomers looked at a color photograph of the sky taken through a large telescope, they observed six yellow stars, two orange stars, a red star and the rest appeared white. The astronomers wanted to determine which of these stars were relatively cool, and which were extremely hot. Arrange the stars they saw in order by color, from coolest to hottest, and explain your sequence.

Experience 9: OBSERVING THE EFFECT OF POSITION ON THE COLOR OF THE SUN

OBJECTIVE

Observe differences in the apparent color of the sun and infer the relationship between the amount of atmosphere the sun's rays pass through and the color of the sun.

PROCEDURE

a. According to today's newspaper, the sun is predicted to rise at ____ a.m. tomorrow. In order to wake up 10 minutes earlier than predicted sunrise, set the alarm clock for____a.m. (Hopefully, there will be a clear sky.)

b. At what time would you expect it to be light?____a.m.

c. Record your sunrise observations of the following:

What time was it when you first observed daylight?___
a.m.

Is this the time you had predicted in b? (Circle *one*:
Yes No)

What color did the sun appear to be when it first rose?

Remember: Never look directly at the sun! This can cause
severe eye damage. Look off to the side of the sun for your
observations.

d. What color did the sun appear to be while you were going
to school? at lunch time? at mid-afternoon?

e. According to the newspaper, the sun is predicted to set at
___p.m. today.

f. At what time is it dark enough that car lights are needed?

g. What color did the sun appear to be just before setting?

h. Diagram 46 shows the sun at four positions in the sky, the
earth, and the atmosphere which extends several hundred miles
beyond the earth's surface. Based on your observations of the
sun, label the color of the sun for the four positions shown.
Using Diagram 47, measure and record the distance the sun's
light must travel through the atmosphere to reach point X on
the earth from points A,B,C and D. (Remember that the
atmosphere does not end abruptly at the line, but a line is
shown here so you can make the necessary measurement.)

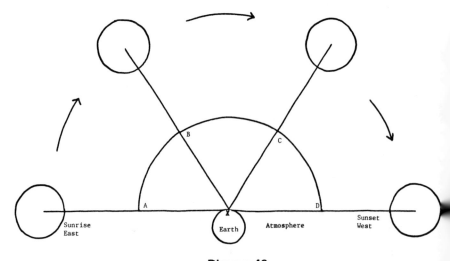

Diagram 46

i. At which times of the day do the sun's rays pass through the greatest distance of atmosphere?

j. What can you infer about the distance the sun's rays travel through the atmosphere and the apparent color of the sun?

MATERIALS

newspaper with sunrise and sunset times
ruler
diagrams

PROCEDURAL INFORMATION

Fewer frustrations will be experienced if sunrise is visible on the morning selected. Careful attention to the weather forecast will help in selecting the day for observing the sun. Spring or Fall would be advisable times to consider for this experience.

It is essential to discuss the dangers to the eyes of looking directly at the sun. The procedure for making the observations should be a very brief glance off to the side of the sun. This should be adequate to determine the approximate color of the sun. Students should be cautioned never to look directly at the sun.

The inference to be drawn is that the greater distance the sun's rays travel through the atmosphere, the more red the sun appears to be. Recall of sunrises and sunsets will reinforce this position.

Colorful sunrises and sunsets have been the subject of poets for many years. More recently, they have become the object of concern of environmentalists as more and more pollution in the skies increases the vividness of the colors. The added effect of

Line	Distance in mm.
A-X	
B-X	
C-X	
D- X	

Diagram 47

pollutants on the sunrise and sunset in many urban areas should be included in the discussion of the experience.

EVALUATION

Shown a picture of a reddish-orange sun, the student will infer the approximate time of day the picture was taken and give an acceptable explanation of the inference.

Experience 10: INFER EARTH FEATURES FROM COLOR RECORDED IN PHOTOGRAPHS

OBJECTIVE

Identify shallow and deep water, rivers, vegetation, clouds, deserts, mountains, peninsulas, deltas, and islands when present in space photographs of the earth and list one or more advantages of using color photographs for interpreting earth features.

PROCEDURE

a. Examine the black and white copy of the photograph of the earth from space. Using a dictionary or atlas, label any of the following features which you are able to recognize:

areas of vegetation	deserts
areas of deep water	man-made objects
areas of shallow water	peninsula
clouds	delta
rivers	islands
mountains or hilly regions	

b. When you have labeled all the features you can recognize, obtain from your teacher a colored photograph of the same scene. Compare your black and white copy with the color photo. If you can now identify more of the features listed above, label them on your map.

c. If there are features present in the photograph which are not on the list, label them and add them to the list. If some of the listed features were not in the photograph, attempt to find them in other photos.

d. What were some of the visual clues you used with the color photograph to identify the features? How do man-made objects differ in appearance from natural features?

e. Were you able to identify more features on the black and white copy than on the color photograph? Explain.

f. Using an atlas, find the names of the countries and bodies of water which are shown in the pictures and label them. According to the scale in the atlas, approximately how large an area do you estimate is shown on the photograph?

g. Based on your observations of the photograph, does the earth appear round or flat? Explain.

VOCABULARY

gulf	atlas
delta	atmosphere
vegetation	peninsula

MATERIALS

several atlases   globe

colored photographs of the earth from space (see books such as *Earth Photographs from Gemini III, IV, and V*, available from the Superintendent of Documents, U.S. Government Printing Office, Washington, D.C. 20402, 1967, $7.00)

black and white copies of colored photographs (Xerox copies)

PROCEDURAL INFORMATION

Color photographs of the earth from space are easily available in back issues of magazines such as *National Geographic* and *Time*. Select a photograph which has good contrast, shows the variety of features you would like to include in the investigation, and shows a familiar area such as the Nile Delta. Black and white Xerox copies of the photograph can be made without removing the original photograph from the magazine or book. If it is not feasible to make a copy for each student, have a group of students do the investigation using tracing paper over their black and white copies for labeling the features. These can be reused by other students. Most likely, several students will have to share the color photograph unless several copies can be found. It would be desirable to use five or six different photographs, giving the students an opportunity to study more than one.

The list of features to be identified in step a, may be altered to meet the needs of the class and to best utilize the photographs. Since the photographs are of a real situation rather than a highly-stylized one, not all desired features will be found in one picture. It is also probable that many additional features will be found as the investigation progresses. These may lead to further extensions of the investigation such as: what direction was the camera facing?; where was the sun?; what is the area in square miles included in the picture?; how can space photos be used to improve maps?; and, what are some uses of photographs such as these?

Color-blind students may experience difficulty in differentiating between some features. If a student is having unusual problems with this activity, you might consider this as a possible cause.

EVALUATION

Given a list of features and several photographs of the earth from space, the student will select five features and identify each in one or more of the photographs.

6

Inquiries into Variations
in Light

Daylight and darkness are with us constantly. The result is that from childhood we come to accept the variations in the length of the daylight and darkness hours for our locality. However, children respond with interest to defining these variations as a reinforcement of their own knowledge. Such reinforcement about their own environment is valuable when discussing variations which occur in other areas such as the North and South Pole, or the Land of the Midnight Sun.

Sunlight has the potential of serving as the solution to our energy requirements. Scientists are seeking efficient ways to convert solar energy into electrical energy and store solar energy so it can be used when the sunlight is not available. Solar batteries, such as those used in these experiences, represent the early progress made in this area. They provide an opportunity for the children to stretch their minds into the future as they find that solar batteries are a part of their world.

Experience 1: DISTINGUISHING CHANGES IN DAYLIGHT
DURING A YEAR

OBJECTIVE

Identify the duration of daylight at the beginning of the four seasons and infer the relationship between the length of daylight and the season.

PROCEDURE

a. Go to a library where newspapers are kept on file and obtain a newspaper from the third week of December, March, June and September. Find the page with the weather facts from each paper and record the sunrise and sunset times from each paper on the chart in Diagram 48.

Month	Date	Sunrise	Sunset	Minutes of Daylight
Dec.				
March				
June				
Sept.				

Diagram 48

b. What do you infer about the length of time of daylight from December to June?

c. What do you infer about the length of time of daylight from June to September? What would you infer about the length of time of daylight from September to the following December?

d. What is the relationship between the length of daylight hours and the warm and cold seasons where you live?

VOCABULARY

daylight   Fall
Summer   Winter
                    Spring

MATERIALS

newspapers from 3rd week of December, March, June, September

PROCEDURAL INFORMATION

Most children can name the seasons in order and recognize that it gets dark earlier as winter approaches. Few have actually verified this information. Such verification as described in the experience can create a feeling of pleasant reassurance in the child if he was unconsciously aware of this previously. It would be appropriate to precede this experience by recording daily sunrise, sunset and length of daylight for a two- or three-week period. A discussion of this record should lead to an inference about where they are in the annual cycle.

EVALUATION

Given the month and day of the year, the student should be able to identify the season and corresponding approximate length of daylight for his community.

Experience 2: DEMONSTRATE A RELATIONSHIP BETWEEN SEASONAL CHANGES AND CHANGES IN DAYLIGHT

OBJECTIVE

Describe and demonstrate, by using a model, the change of seasons as the earth revolves around the sun and the accompanying changes in daylight and darkness.

PROCEDURE

a. Measure the diameter of the globe and cut a hole of that size diameter in the piece of oak tag.

b. Place the globe so its North Pole is pointing toward the North Pole of the earth and Polaris.

c. Place the light bulb that represents the sun a few feet from the globe. Rotate the globe. What season of the year does this represent from where you live?

d. Hold the cardboard on the globe so that it is perpendicular to the floor with one side facing the sun. The cardboard should divide the globe into two equal halves, one which is light and one which is dark. See Diagram 49.

e. Observe the part of the globe on the side of the cardboard toward the sun. This represents the daylight side of the earth.

f. Mark a spot on the equator. As the earth rotates, during what part of the rotation is that spot found on the light side of the cardboard? If one rotation takes 24 hours, how long would the spot be in light at the equator?

g. Mark a spot at 40 degrees north of the equator. When the globe is rotated this time, during what part of the rotation is the spot on the light side of the cardboard? How long would the spot be in the light at 40 degrees N.? Select a spot near the North Pole. How long would the spot be in the light during one rotation?

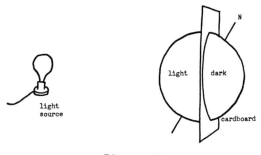

light source

Diagram 49

h. What season would you infer it to be where you live according to your model?

i. How is the southern hemisphere position different from the northern hemisphere?

j. Being careful to keep the North Pole pointing toward the same spot, move the globe counterclockwise through one-fourth of a circle as shown in Diagram 50. Rotate the cardboard so it again divides the globe into the light and dark side.

k. On the basis of this new position, repeat steps f, g, h, and i. Record the results.

l. Again move the globe counterclockwise through one-fourth of a circle, reposition the cardboard, and repeat steps f, g, h, and i

m. Repeat step l.

n. From what you have observed, what can you tell about the number of daylight hours and the seasons and how these change?

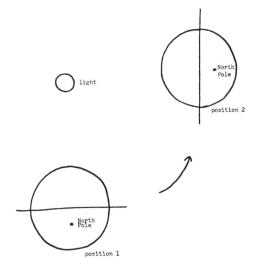

Diagram 50

VOCABULARY

day   perpendicular
night   axis
Polaris   inclination

MATERIALS

globe, one per group
oak tag (larger than the globe's diameter) one per group
light bulb and socket, one per group

PROCEDURAL INFORMATION

It is important that the axis of the globe is pointed toward the same place throughout the activity. It is advisable that this place be a spot or an imagined spot on the north wall. Perhaps a star to indicate Polaris could be placed on the north wall of the classroom at an angle equal to the latitude of the school. This would permit the axis of the globe to be in line with Polaris at all four seasonal positions. Globes of 8″ diameter or larger are best to use.

EVALUATION

Given two relative positions of the earth and sun, using the model, the student will identify the season for a location of his choice and explain the differences in daylight and darkness in each case.

Experience 3: CONSTRUCTING A LIGHT METER

OBJECTIVE

Construct a light meter using a photoelectric cell and demonstrate that it is operational.

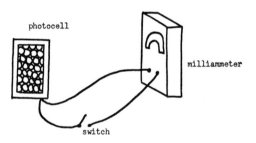

photocell

milliammeter

switch

Diagram 51

PROCEDURE

a. Fasten the milliammeter and switch to the board. A milliammeter is a very sensitive instrument which measures the presence of an electric current. See Diagram 51.

b. Attach the photoelectric cell (solar battery) on the same board in an upright position.

c. Connect the photoelectric cell, milliammeter, and switch in a series circuit. Place the board so the sunlight falls on the cell and close the switch. If the meter needle moves farther towards zero, reverse the wires from the meter to the photoelectric cell and switch. When operational, the milliamp reading should correspond with the MA given on the cell or in the accompanying literature.

VOCABULARY

milliammeter	solar battery
milliampere	complete circuit
photoelectric cell	series circuit

MATERIALS

photoelectric cell (about $2.00 from a radio store or from: Allied Radio, Chicago 80, Ill.; Radio Shack, Boston 17, Mass.; Edmund Scientific Co., Barrington, N.J.)
milliammeter (range of 0-25 or 0-50 milliamps)
off-on switch   screws or nails
board   bell wire
            screwdriver or hammer

PROCEDURAL INFORMATION

The photoelectric cell or solar battery is able to convert light energy into electrical energy. If this is the first experience the students have had with the cells, it would be advisable to provide ample time for examination, research, and discussion. If sunlight is not available for determining if the circuit is operational, an electric light can be used as the light source. The students should be cautioned about excessive heat from the light and instructed to keep the cell about 10 inches or more from the light. Precautions about rough handling should also be given.

The meter and cell combination can be compared with the light meter of a camera. The use of solar cells in satellites as well as their potential on earth may be of interest for some students as extensions of the experiences.

EVALUATION

Given a photoelectric cell, switch, milliammeter, and wires, the student will be able to check the circuit to determine if the system is properly connected and describe how it should respond if turned on.

Experience 4: MEASURING CHANGES IN LIGHT INTENSITY

OBJECTIVE

Infer the relationship between the distance from a light source and the electrical output observed on the light meter.

LIGHT SOURCE	1st DISTANCE FROM LIGHT METER	MILLIAMPS	2nd DISTANCE FROM LIGHT METER	MILLIAMPS
25 w bulb				
75 w bulb				
100 w bulb				
150 w bulb				

Diagram 52

PROCEDURE

a. Position the light meter at one end of a meter stick and a 25-watt incandescent light at the other end of the stick. Darken the room, turn on the light and the light meter. Record the reading on the chart in Diagram 52. Move the light to 50 cm. from the light meter. Again, record the result.

b. Repeat this procedure for the other light sources and record the results.

c. On the basis of your data, what can you infer about the different light sources?

d. What is the relationship between the distance from the light source and the amount of illumination recorded by the light meter? If in doubt, check this by comparing the readings from the same light source at 100 cm., 50 cm. and 25 cm. from the meter.

e. On the basis of your inferences, what would you predict the readings from the sun to be like at sunrise, noon and sunset on a sunny day? Test your prediction. Describe your prediction in terms of your test results.

VOCABULARY

incandescent light   illumination

MATERIALS

light meter (as constructed or a commercial one)
meter stick
portable light socket
4-5 different-wattage light bulbs

PROCEDURAL INFORMATION

The students should infer that different-watt light bulbs produce different amounts of illumination. They should also infer that as a light source is moved closer to the photoelectric cell the illumination increases, and as the source moves farther away the illumination decreases. The illumination increases by the square of the distance as the distance decreases, and decreases by the square of the distance as the distance increases.

The students may reason that the reading from the sun should be the same for all three times since the distance remains the same in all three cases. A lower reading at sunrise and sunset should be expected since the diffusion at these times is greater.

EVALUATION

Given the following information, what would you infer about the distance between the source and the cell?

source	milliamps
100 W	5
100 W	8
100 W	12

Experience 5: MEASURING LIGHT INTENSITY OF DIFFERENT REFLECTIVE SURFACES

OBJECTIVE

Record the light meter readings, as light is reflected from various color surfaces, and index the colors from most reflective to least reflective.

PROCEDURE

a. Place a 100-watt bulb in a socket with a reflector and place it 50 cm. from a sheet of colored paper. Place the light meter 25 cm. from the paper so that a line from the bulb to the paper would be reflected at a 90° angle to the meter.

b. Darken the room and, if necessary, reposition the light meter either closer or farther from the paper until you get a reading which places the needle near the middle of the scale. Be sure the light meter is only recording the light reflected from the colored paper. Record the light-meter reading on the chart in Diagram 53.

Paper color	Distance (cm.) from light to paper	Distance (cm.) from paper to light meter	Milliamps
Green	50		
Red			
Tan			
Yellow			
Black			
White			
Purple			

Diagram 53

c. Change the color of paper and record the results. Be sure not to change the position of the light meter. Complete the readings for the remainder of colors being used.

d. Index the colors on the basis of the highest to the lowest meter readings.

e. Which colors seem to be the best reflectors? The poorest?

f. On a sunny day, take the light meter outdoors. Look around you and identify those surfaces which you feel would be good reflectors and those which would not be good reflectors (e.g. grass, dirt, buildings, windows, blacktop). Position the light meter so it is a known distance from one of the surfaces and record the reading. Repeat this for the other surfaces. Discuss any errors in your predictions with the class.

VOCABULARY

light intensity   reflector
              90-degree angle

MATERIAL

light meter   meter stick
100-watt incandescent light bulb with socket and reflector
colored paper with same texture and luster, red, yellow, green, tan, white, black, purple

PROCEDURAL INFORMATION

The distance between the paper and light source and the paper and meter should remain constant throughout the activity. Also, the texture and luster of the different colors of paper should be the same. A discussion of why there are differences in the light reflected from different color surfaces outdoors may be partially answered by touching the surfaces to compare their relative temperature. The light which is not reflected is converted to heat energy making the dark colored surfaces warmer.

EVALUATION

Given four pieces of paper different only in color, the student will index them on the basis of most reflective to least reflective surfaces.

7

What to Do with Temperature Changes

The temperature is perhaps one of the best topics one can use to provide children with experiences in the processes of science. Children are interested in the temperature as it influences their dress and activities. They are, also, interested in their own body temperature. To use the temperature as a means of providing children with experiences in the processes of science makes such experiences high interest activities.

The deluge of information and the constant need for updating knowledge emphasize the need for children to be able to use the skills or processes of science. Temperature and temperature change are highly versatile as children observe and measure the temperature; record and graph temperature changes; predict future temperatures and evaluate their predictions; and state hypotheses regarding rates of change and conduct experiments based on their hypotheses.

The objectives for the experiences involve the children in several graphing experiences and interpreting their information from the graphs. Cycles and changes become more easily identified from their graphs than from lists of figures. These experiences serve as a starting point for the creative teacher and students as they relate the use of this process of science to other areas of the curriculum.

Experience 1: CONSTRUCTING A THERMOGRAPH OF
THE CLASSROOM

OBJECTIVE

Record temperature variations and use them to construct a thermograph of the classroom.

PROCEDURE

a. Select at least 10 locations which are at the same distance above the classroom floor. Select sites around the room such as near the windows, near the center of the room, and near the door. Measure and record the temperature at each of these sites.

b. Draw a scale map of the classroom showing the locations which were selected and the temperature at each.

c. Draw a line which connects each of the locations having the same temperature. These lines represent areas of similar temperature and are called isotherms. The map showing the isotherms is called a thermograph.

d. Repeat steps a to c but take all of the temperature readings at a different distance above the floor. Compare your results.

e. Compare your thermographs. Are the temperatures the same for both levels?

f. Would you expect a thermograph made at the floor level of a room to be the same as one made at the ceiling of the same room? What effect would windows or air vents have on the temperatures near them?

g. If you were to make a thermograph outdoors, what do you think it would be like?

h. Obtain a copy of a weather map from the newspaper. Observe that there are isotherms shown on the map. What can you tell about the temperature from this map?

i. What information does the weather bureau need to make a thermograph of the United States? Find out how they obtain their temperature readings.

VOCABULARY

temperature   isotherm   thermograph

MATERIALS

> several thermometers   meter stick
>   large sheets of newsprint

PROCEDURAL INFORMATION

The construction of a thermograph is similar to drawing contours on a contour map. The students record the differences in temperature for similar elevations in the room. The more readings which can be taken at a given elevation, the more participation and interest will be generated. The students may want to discuss or explain the selection of the locations.

Comparisons of at least three thermographs, one near the floor, one at seat level, and one near the ceiling should reveal air flow patterns in the room. If the room has a cooling or heating system, the readings should all be taken either when it is on or when it is off. A discussion of the thermographs may reveal drafty areas in the room which can be confirmed by the children seated in those areas.

Weather permitting, it may be more desirable to conduct the entire experience out-of-doors on a section of the school grounds. Try to include shaded and sunny areas as well as areas protected by a building or near a sidewalk or blacktop.

Relating the thermographs to a broader situation can be done using the weather maps from the newspaper. Research will reveal a vast network of data-collecting stations throughout the United States. These stations report their findings continually to the U.S. Weather Bureau for compilation of weather charts which include temperature, pressure, precipitation, and wind direction. This experience could serve to stimulate interest in other weather factors on the part of the students.

EVALUATION

Given a map and a listing of temperatures for given sites on the map, the student will construct a thermograph.

Experience 2: GRAPHING TEMPERATURE CHANGES AND INFERRING CAUSES

OBJECTIVE

Record on a graph the temperature changes during a 24-hour period of time and infer causes of the change.

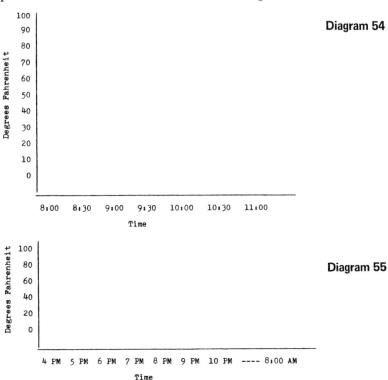

Diagram 54

Diagram 55

PROCEDURE

a. Place a thermometer outdoors so that it will not be in the sun and is easily visible. Record the temperature every half-hour for the school day.

b. Construct a bar graph, showing the temperatures recorded by coloring in the columns to the level equal to the degrees recorded. See Diagram 54.

c. Using your graph, can you infer what the temperature was 15 minutes before or after the hour? How accurate is your inference?

d. Using your graph, can you predict what the temperature will be one hour after school? Two hours after school? How accurate is your prediction?

e. Predict what you think will happen to the temperature during the night. Check your prediction in the newspaper weather reports on the hourly temperature record. Graph the data from the newspaper from 4:00 p.m. until 8:00 a.m. of the night selected (Diagram 55) and compare the two graphs you have made.

f. What was the recorded high? At what time did it occur? What was the overnight low? At what time did it occur?

g. What can you infer were the causes of the changes which occurred in the temperature?

VOCABULARY

temperature	Fahrenheit
thermometer	scale
Celsius	

MATERIALS

graph paper   thermometers
newspaper having local weather data

PROCEDURAL INFORMATION

During the graphing exercises, the student prepares a bar graph in the first part of the activity. Later, there is an opportunity for another graph to be prepared. A line graph may be preferred to provide variety in the graphing experience.

The newspaper records the temperature in degrees Fahrenheit. To provide experience in using the metric system, it may be desirable to record and graph the temperature for the day in both Fahrenheit and Celsius. Comparisons of such temperatures as body temperature, average room temperature, and the freezing and boiling points of water should be made between the two scales. Attempts to convert temperatures from one scale to the other are difficult, unnecessary, and should be avoided at this time.

EVALUATION

Given a graph showing temperature variations for a six hour period of time, the student will infer if the time shown is a.m. or p.m. and state the observations and inferences supporting his choice.

Experience 3: IDENTIFYING PATTERNS IN TEMPERATURE CHANGES DURING 7 DAYS

OBJECTIVE

Graph the daily temperatures for a 7-day period and identify any patterns which appear in the temperature changes.

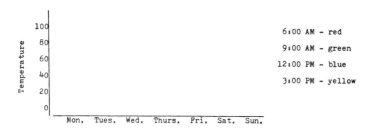

Diagram 56

PROCEDURE

a. Place a thermometer out-of-doors so it can be easily read but is not in the direct sunlight. Record the temperature at 9:00 a.m., 12:00 m. and 3:00 p.m.

b. Prepare a graph each day showing the data obtained either from direct readings or the newspaper for 6:00 p.m., 9:00 p.m., 12:00 p.m., 3:00 a.m. and 6:00 a.m. Be sure to use the same scale for each graph so they can be easily compared.

c. Are there any patterns which appear in your graphs? Describe any patterns you find.

d. Construct a line graph which shows the temperature recorded daily at 6:00 a.m. with a red line, at 9:00 a.m. with a green line, at 12:00 m. with a blue line, and 3:00 p.m. with a yellow line. Label the axis as shown in Diagram 56.

e. Does the graph in d show any patterns which were not seen before? Describe them.

f. You may wish to make another graph like the one in d to show the temperatures for the other times recorded.

g. When does the warmest time of the day usually occur?
   When is the coolest time of the day?
   Can you infer the temperature for 8:00 a.m.?
   2:00 p.m.?

VOCABULARY

patterns

MATERIALS

thermometers	newspaper with weather data
graph paper	colored pencils

PROCEDURAL INFORMATION

The range of temperature changes will vary with the season. You may want to have the students infer how their graphs would appear if they had been done during other seasons. They would then consider such things as: how would the temperature range compare with the one just completed?; would the daily cycle still occur in the other seasons?; would there be any differences in the relationship between the time and temperature in the different seasons?

The students may be given the option of collecting their own data, obtaining all data from the newspaper, or something between these two. In any case, the data will be graphed on a daily basis. Again an option, this time for the teacher, as to whether each student will prepare a complete set of graphs, one complete set will be prepared by the class as a whole, or something between the two. A similar option is open for the graph for part d showing the temperature for a given time each day. The choices made may be dependent upon the students' need for graphing experiences.

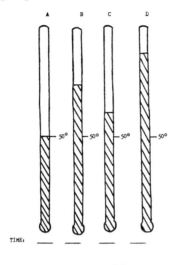

Diagram 57

EVALUATION

a. The readings on the thermometers in Diagram 57 were taken at 4:00 p.m., 9:00 p.m., 12:00 p.m. and 6:00 a.m. of the same clear day. Label each thermometer by the time you infer it would have been, based on the reading.

Experience 4: MEASURING WATER TEMPERATURE AT VARIOUS DEPTHS

OBJECTIVE

Construct a sampling bottle which will provide temperature readings for samples taken at various depths; graph the water temperature of samples from different depths; and infer the causes of differences or similarities.

PROCEDURE

a. Construct a sampling device such as that shown in Diagram 58 (or design and construct one of your own). Be sure that all connections are air tight between the stopper, bottle, thermometer, and tubing, and that the weight is securely fastened to the bottle.

b. Securely attach a rope to the sampler and, starting at the sampler, tie a knot in the rope every two feet above the sampler.

c. Collect a surface sample and record the temperature of the sample. Collect a sample from 2' below the surface and record the temperature. To collect a sample, lower the sampling device to the desired depth and open the pinch clamp on the rubber tube. Be sure to empty the bottle and securely replace the cover and pinch the clamp before collecting the next sample. Repeat this procedure at least for samples from 4', 6' and 8'.

d. Why wouldn't you just lower the thermometer and then pull it up quickly to get the reading?

e. Graph your data showing the relationship between depth of water and temperature.

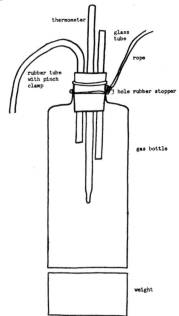

Diagram 58

f. Does the temperature change with the depth? If so, in what way?

g. Would you expect to get similar results from other areas of the same body of water? Select another location and repeat the sampling process.

h. Graph the data from the second location. How do you explain the two graphs?

VOCABULARY

sampling bottle

MATERIALS

wide mouth bottle with 3-hole rubber stopper
several feet of rubber tubing
several feet of rope
thermometer   pinch clamp
glass tubing   weight

PROCEDURAL INFORMATION

The subject of oceanography is appealing to students, but they have little comprehension of the problems confronting the oceanographer, even when he attempts something as simple as finding the temperature of water at different depths. Several variations can be made in obtaining the water temperature, including electronic devices such as a thermistor. The students may wish to pursue further how oceanographers resolve this problem and the reasons they have for obtaining the temperature in the first place to make the experience more relevant.

There are several alternatives which students may suggest in making the sampling device. If you provide them with this option, it is essential that they understand the need for obtaining the water sample from each level. Some questions about the samples which should be discussed include: why doesn't the water go into the bottle before opening the clamp?; why do we need a weight?; how heavy does the weight have to be? In addition, the class should arrive at a standard operational procedure which can be used for obtaining all samples and readings.

Samples can be obtained from a pond, lake or even a swimming pool. Temperatures will vary with such things as depth, water circulation, time of day, season and turbidity. The influence of these factors should be discussed. If students are interested, they may want to test some of their inferences regarding these variables.

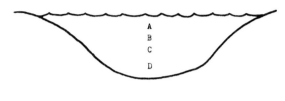

Diagram 59

EVALUATION

If the water in a lake is divided into four levels (level A the top layer, B the next lower level, and on to D the lowest level), infer which level would be coldest in the summer and explain your answer. See Diagram 59.

Experience 5: IDENTIFYING COOLING RATES OF DIFFERENT GROUND COVERS

OBJECTIVE

Observe and graph temperature changes in different ground covers and sequence them from fastest-cooling to slowest-cooling.

PROCEDURE

a. Select a section of ground covered by grass which is in the sunshine. Place a thermometer in the grass and record the temperature every 1/2 hour, starting at 4:00 p.m. and continuing until darkness.

b. Select several other types of ground cover such as

blacktop, dirt, or cement and repeat step a. Since you will have several thermometers, be sure to keep careful records of each temperature change.

c. Observe each surface where there is a thermometer and record a description of the color and texture of each along with the temperatures. Predict which surface will cool the fastest and which will cool the slowest.

d. Make a graph which shows the temperature changes of each of the thermometers. Compare the graphs and on the basis of the information sequence the surfaces in order, starting with the one which cooled the fastest and on through to the one which cooled the slowest.

e. Check your results with your predictions. How do you explain these?

f. Is there any relationship between the color of the surface and its cooling rate?

g. Is there any relationship between the texture of the surface and its cooling rate?

VOCABULARY

cooling rate	texture
temperature	radiation

MATERIALS

graph paper	thermometers

PROCEDURAL INFORMATION

This activity can be conducted at any of several different times. The cooling rate which accompanies sunset is used here. It is possible to use the warming rate, starting with early morning and continuing on until around 11:00 a.m. If neither of these is convenient, the activity could be done in a shorter time period, selecting a sunny area which will be in a shadow shortly or a shaded area which will soon be sunny. The temperature range will. be less but the time will, also, be less.

A discussion of radiant energy and the advantages of selecting heat-reflective and absorbent surfaces may be part of the follow-up of the activities. These could include the advantage of a blacktop driveway in the winter and the disadvantage of the same surface in the summer, or the effect a large, blacktop parking lot would have on its surroundings.

EVALUATION

Given a graph showing the cooling rates of several surfaces, identify the surface which cools the fastest and the one which cools the slowest.

8

Investigating Changes of Form

One of the early experiences of young children living in northern latitudes involves the change of form which occurs when they first try to save that snowball or icicle. The resulting puddle of water provides a very real introduction to the concept of melting. Although children are familiar with changes in form, they are generally unaware of the accompanying factors which cause or accompany these changes.

Experiences are meaningful that involve children in observing changes of form, inferring possible causes of such changes as well as in recording and graphing data obtained when these changes occur. Relating change of form to metamorphic and igneous rocks is done more easily than relating it to sedimentary rocks. However, the cementing which occurred when cementing materials found their way into some sediments can be associated with changes in form.

Experience 1: IDENTIFYING DIFFERENCES IN FORM

OBJECTIVE

Observe solids, liquids, and gases and identify the different characteristics of each.

PROCEDURE

a. Select a balloon and describe its shape. Inflate the balloon, tie it, and describe its shape.

b. Select another balloon and observe its shape. Partially fill it with water, tie it, and describe its shape.

c. Select a third balloon and partially fill it with pebbles. Describe its shape.

d. One balloon is filled with gases, one with a liquid, and one with a solid. Observe each of these. You cannot see the contents of any of the balloons but you can observe some of their characteristics. Make a list of the characteristics of a liquid, a gas, and a solid, based on your observations.

e. Place several ice cubes in a glass and make a diagram of the glass and ice. Allow the ice to melt. What difference do you observe in the ice and the water?

f. Hold your nose and remove the top from the jar of moth balls. Hold it a foot away from your nose. Can you smell the moth balls? Why not? Let go of your nose. Can you smell the moth balls? Why?

g. Add to your list in d any additional characteristics you have observed.

VOCABULARY

solid	gas
liquid	matter
forms of matter	

MATERIALS

balloons	glass
water	ice cubes
pebbles	jar of moth balls

PROCEDURAL INFORMATION

Most children can tell whether a thing is a solid or a liquid. They have had many experiences with these two forms of matter. On the basis of these experiences and their observations, they are able to identify the differences between solids and liquids. To identify the characteristics of the invisible third form of matter, even though they know it is there, is difficult for some students.

The fact that a solid has a shape of its own, compared with a liquid which takes the shape of the container it is in, can evolve from students' observations. To differentiate between a liquid and a gas, however, may well require careful questioning by the teacher.

EVALUATION

Given a variety of solids, liquids, and gases, group them on the basis of being a solid, liquid, or gas and explain your decisions.

Experience 2: OBSERVING CHANGES OF FORM

OBJECTIVE

Observe changes in form and infer the cause of each change.

PROCEDURE

a. Place a small piece of ice into a test tube and hold the test tube in a beaker of hot water. Observe and record what happens to the ice.

b. Place a small piece of ice into another test tube. Using a test tube holder, hold the test tube over a source of heat. (Be sure to hold the test tube at a slight angle with the mouth pointed away from you and other people.) Hold a glass of cold water over the mouth of this test tube. Observe what happens to the ice and record your observations.

c. Half-fill a small beaker with pieces of ice and place it into a larger beaker. Pour water into the larger beaker until it is about an inch deep around the smaller beaker. Pour several teaspoonsful of salt onto the ice. Stir the salt and ice with a spoon or stick. Observe and record what happens on the outside of both the smaller beaker and the larger beaker.

d. Infer the cause for the changes you have seen in each of these experiences.

VOCABULARY

solid	evaporate
gas	freeze
liquid	melt
condense	sublime

MATERIALS

ice	beakers (2 sizes)
test tube	salt
test tube holder	spoon
alcohol burner	glass

PROCEDURAL INFORMATION

When the ice is placed in warm water, it will melt slowly. When the ice is placed in a hot flame, it will change from a solid directly to a gas. This process, called sublimation, occurs when enough heat is present. This could also be achieved when ice cubes are placed in a hot sauce pan or an electric hot plate or in an electric fry pan.

The condensation which appears on the outside of the glass of cold water in step b is an indication that the solid (ice) changed to a gas (water vapor) and then to a liquid.

An interesting and tasty application of the salt and ice combination is a demonstration of how ice cream used to be made. The salt and snow or ice were placed in the bucket of the freezer around the cylinder containing the ice cream mix. As the ice cream was stirred, the salt lowered the freezing point of the snow around it. The snow and salt lowered the temperature of the ice cream until it hardened.

EVALUATION

Given a solid and a liquid, explain what would be necessary to change the solid to the liquid and the liquid to the solid.

Experience 3: INFER THE CAUSE OF CHANGE IN MOTH BALLS

OBJECTIVE

Observe changes in form in moth balls and infer the causes of the changes.

PROCEDURE

a. Place a small moth ball in a test tube and place the test tube into a beaker of hot water. Observe the moth ball and record your observation.

b. Observe as your teacher heats another moth ball in a test tube. What happens to the size of the moth ball? Observe the jar at the mouth of the test tube. Where did the material come from that appears in the jar? On what observations do you base your answer?

c. What can you infer were the causes of the changes you have observed?

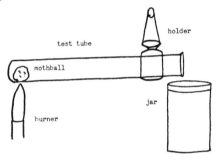

Diagram 60

VOCABULARY

evaporation   condensation

sublimation

MATERIALS

moth balls   test tube holder
test tubes   jar
hot water   Bunsen burner (or propane burner)

PROCEDURAL INFORMATION

The teacher does the activity in b while the class observes or a student may perform the activity for small groups. See Diagram 60. In either case, the following should be tried before it is done in front of the group. As the moth ball is heated, it will melt and evaporate. By holding the test tube nearly horizontal, the "moth ball gas" can be poured into the cooler jar leaving the liquid in the test tube. The gas changes directly to a solid (sublimates) in the jar. The students should infer that heat was required to vaporize the moth ball and that sudden cooling caused the sublimation. The drastic temperature change resulted in sublimation rather than condensation.

It is advisable to conduct this activity in a well ventilated room.

EVALUATION

Given a piece of paraffin, infer what would be necessary to change it to a liquid and back to a solid.

Experience 4: GRAPHING TEMPERATURE CHANGES ACCOMPANYING CHANGES IN FORM

OBJECTIVE

Observe a solid change to a liquid and a liquid change to a gas, record your observations of temperature changes on a graph, and state inferences regarding changes of form based on your observations.

PROCEDURE

a. Fill a cup about 3/4 full of crushed ice and cover the ice with water. Place a thermometer in the mixture. Describe what you think will happen to the temperature of the ice water mixture.

b. Stir the mixture constantly and record your observations of the ice water every 5 minutes on the chart in Diagram 61.

Time	Temperature	State of Matter		
	in degrees C	Solid	Liquid	Gas

Diagram 61

Continue to record your observations at least 4 times after the ice has melted.

c. What is the temperature of the room in which you are doing this activity?

d. How warm do you think the ice will get?

e. Where do you think the heat comes from to melt the ice and warm the water?

f. Make a line graph of your data showing temperature on one axis and time in minutes on the other. Compare your graph with your description of what you thought would happen in step a.

g. Compare your results with others in the class.

h. Place a beaker of water on a hot plate and record the temperature of the water. How hot do you think the water will get? Turn on the hot plate and record the temperature each minute on a chart like the one used in b.

i. What happened to the temperature when the water began to boil?

j. What change of form is occurring?

k. After the water starts boiling, take at least 5 more readings. Graph your data.

l. How does your data compare with your prediction? How do you explain this?

m. How do your results compare with the rest of the class?

n. How do you explain the temperature remaining the same for so long in both of these activities?

VOCABULARY

melt	Celsius
Fahrenheit	boil

MATERIALS

crushed ice   hot plate
beaker   thermometer

PROCEDURAL INFORMATION

When describing what will happen to the temperature of the ice water, most students will expect a steady increase in the temperature. Similarly, most will expect not only a steady rise in the temperature of the water, but also a much higher temperature than is obtained as the water is heated. The temperature remains the same in both cases while the heat being added provides the necessary energy to effect the change of form. In the case of the ice, the temperature rises after the change from ice to water is completed. The hot water changes to steam as heat is added, so by the time the change from a liquid to a gas is completed the water is gone.

It is important that the ice water be well stirred each time before taking the temperature reading. The stirring results in a more even distribution of the heat. When reading the temperature of the hot water, be sure that the thermometer is not resting directly on the bottom of the beaker. This could result in inaccurate readings.

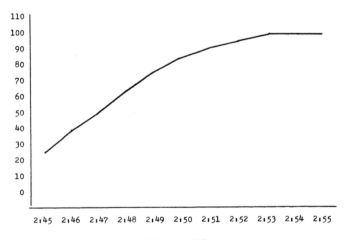

Diagram 62

EVALUATION

Given the graph in Diagram 62, explain why the temperature remained the same for the last 3 readings.

Experience 5: DEMONSTRATING CEMENTATION BY CHANGE
OF FORM

OBJECTIVE

Construct a model to show the cementing of rock material as a change of form takes place.

PROCEDURE

a. Spread a thin layer of fine gravel in the bottom of a dish about 8 cm. high and 8-10 cm. in diameter. Stand a candle which is about 2-3 cm. in diameter in the center of the dish so that the candle is the same height as the top of the dish. Fill the remainder of the dish with gravel and light the candle.

b. Place the dish near an open window or where the flame can get plenty of air for burning. When the candle has completely burned, let the dish cool. Then tip the dish upside down and carefully remove the gravel.

c. Examine what came out of the dish. How does it differ from the gravel you originally put in the dish?

d. What is holding the gravel particles together? How do you think it got there?

e. What is left where the candle was? Could this be used as a mold to make more candles?

VOCABULARY

cementing   cementation
                molten

MATERIALS

dish (about 8 cm. in diameter and 10 cm. high)
candle (2-3 cm. in diameter and 10 cm. high)
aquarium gravel
matches

PROCEDURAL INFORMATION

An inexpensive candle which drips as it burns is essential for this activity. As the candle burns the gravel particles become very hot. Some of the wax is burned. Some of it melts and spreads through the gravel binding it together as it cools. This very fine layer of wax has an amazing binding force. There are many earth materials which act as cementing materials as the wax did.

As the candle burned, it left behind a hole. This hole could serve as a cast from which one could mold another candle. This cast can be compared with the casts made hundreds of years ago as decaying animals on a beach were covered with sediment. The air spaces formed cast-like areas which were slowly filled in by materials being deposited, creating fossils. An interesting example of replacement by dissolved materials is the replacement of the organic matter in buried logs to produce petrified wood.

The students should be cautioned about handling the jar or its contents after the candle starts burning. Also, be sure that there is a layer of gravel between the inside of the dish and the bottom of the candle.

EVALUATION

Given small particles of sand, describe a process of cementing them together.

Experience 6: CONSTRUCTING SEDIMENTARY ROCKS BY CEMENTATION

OBJECTIVE

Construct a model to show how fine sediments can be cemented and infer a relationship to the formation of rocks.

PROCEDURE

a. Make pencil-point sized holes all over in a styrofoam cup including the bottom. Fill the cup with coarse sand.

b. Make a mixture of half white liquid glue and half water and pour it into the cup of sand. Be sure and hold a second cup underneath the first one to catch the extra glue. See Diagram 63. Pour the extra glue you catch back through the cup of sand 5 or 6 times.

c. Put the cup of sand aside to dry for a day.

d. When the sand has dried, break the styrofoam cup away from the sand. Compare what you have with some of the original sand.

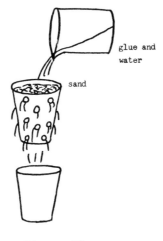

glue and water

sand

Diagram 63

e. Why are the particles of sand holding together?

f. Do you think think a similar process could happen in the earth's crust to form rocks? Where would the glue come from?

g. How solid is your rock? Will it weather very fast?

VOCABULARY

sedimentary rock   weather

MATERIALS

styrofoam cups   white liquid glue
pencil   water
                    sand

PROCEDURAL INFORMATION

There are many natural cementing materials in the earth's crust. As they are dissolved from surface rocks, they may eventually fill in the spaces between particles of sand and gravel, acting as a binding agent. As the cementing materials are deposited, sedimentary rocks such as sandstones and conglomerates are formed.

The glue which is used is water-soluble. The newly made rock is also water-soluble. It can be used to demonstrate weathering as the glue is dissolved away from the homemade sandstone.

EVALUATION

Given a piece of basalt, conglomerate and granite, select the sedimentary rock and explain the basis for your selection.

Experience 7: INFER THE FORMATION OF SEDIMENTARY ROCKS FROM PRECIPITATES

OBJECTIVE

Describe a precipitate and the relationship between precipitates and sedimentary rocks.

PROCEDURE

a. Slowly add salt to a jar of warm water. Stir the mixture and continue adding salt until no more will dissolve.

b. Repeat this process to form a mixture of gypsum and one of calcium carbonate. Label each container.

c. Place the jars on a table where you can observe the contents of each jar, once a day, without disturbing the mixtures. What do you think will happen to the contents of each jar?

d. Compare the contents of each jar after the water has evaporated. How are they alike? How are they different?

e. Using a dictionary, find the definition of a precipitate.

f. Ocean water has many minerals dissolved in it. What do you think would happen to these minerals if the ocean water in a bay were to evaporate? Can you find any examples of this sort of thing having happened in the history of the earth?

g. What is a sedimentary rock? Use a science dictionary if one is available.

h. What do precipitates have to do with sedimentary rocks?

VOCABULARY

gypsum   precipitate
calcium carbonate   sediment
minerals   sedimentary rocks

MATERIALS

3 jars of warm water   spoon
gypsum (powdered)   table salt
calcium carbonate   dictionary
    (powdered)

PROCEDURAL INFORMATION

If possible, it would be of interest to the students to be able to examine samples of rock salt, gypsum (or alabaster, a variety of gypsum), and calcite (calcium carbonate). A discussion of the salt deposited at Great Salt Lake and the many salt wells and mines in the United States may add relevance to this activity. One of the largest salt mines of the world is located in Retsof, New York. Similar discussions of other mineral sediments may be used to relate this experience with economics, history and geography.

A field trip to available salt, gypsum, or other mining operations in the area would be advantageous. However, plans should be made well in advance. Safety precautions may prohibit students from entering the mines but much can be learned from surface operations.

EVALUATION

Tell what a precipitate is and what connection precipitates have with sedimentary rocks.

9

How Environment Affects
Magnetic Fields

Give most children or adults two magnets and the results are quite similar. No matter how familiar one is with magnets, the first reaction is to investigate or re-investigate what happens as the two are brought together. One might infer that the manufacturers of children's toys recognize that magnets contribute a high degree of fascination to games.

The reactions of opposite magnetic fields can be felt by the person manipulating the magnets. The tactile and visual observation of the interactions between magnets provides students with an energy field which can be explored in complete safety. Such an advantage is not afforded all forms of energy. The accompanying experiences for the students present opportunities to observe and identify the characteristics of magnets, the interactions between magnets, and the interactions between magnets and their surroundings.

Experience 1: OBSERVING THE EFFECT OF MAGNETISM ON DIFFERENT OBJECTS

OBJECTIVE

Observe magnets and describe their effect on objects.

PROCEDURE

a. Place a bar magnet on top of a pile of books so that one end of the magnet extends out over one side of the pile of books.

b. Tie a paper clip to one end of a three-foot piece of thread. Let the paper clip come in contact with the end of the magnet which overhangs the books.

c. Place the loose end of the thread under another book on the table about 18 inches away from the pile.

d. Slowly pull on the end of the thread, keeping it under the book. As you pull the thread between the paper clip and the book, it becomes tighter. Continue pulling until the paper clip is pulled from the magnet about 1/4 to 1/2 inch. This is tricky but the paper clip can be suspended in mid-air as shown in Diagram 64.

e. Place various objects between the magnet and paper clip, recording your observations on the chart in Diagram 65. Add other objects to your list.

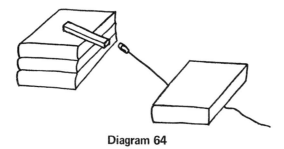

Diagram 64

Material	Observation
paper	
glass	
wood	
nail	
cardboard	
paper clip	
cloth	

Diagram 65

VOCABULARY

    magnet   magnetic
        non-magnetic

MATERIALS

bar magnet	thread
books	paper
paper clips (or needle)	cardboard
wood	glass
nail (iron)	cloth

PROCEDURAL INFORMATION

At first, the suspended paper clip will fascinate the students. They should leave time to try such things as who can get the paper clip farthest from the magnet before it drops. After they have placed a variety of objects between the paper clip and magnet, they should be able to describe the difference between the objects which do affect the paper clip and those which don't.

EVALUATION

Given a list of various objects, a magnet, a paper clip, and thread, identify which objects will affect the paper clips, either from recall or by testing them.

Experience 2: IDENTIFY THE PROPERTIES OF MAGNETS

OBJECTIVE

Observe magnets and describe the effect on each other.

PROCEDURE

a. Suspend a bar magnet on a string above a table top as shown in Diagram 66. Draw a straight line on a piece of paper and place it below the magnet so the line is lined up with the magnet. Label the ends of the paper the same as the ends of the magnet.

b. Rotate the magnet half way around a circle, release it and wait until it stops moving. Observe the position of the magnet. Did the magnet return to the same position? Repeat this several times and describe your results.

c. Place one end of another bar magnet first near one end of the suspended magnet, and then the other. Describe the result in each case.

d. Hold the opposite end of the second magnet near one end of the suspended magnet, then the other. Describe your results.

e. Describe how magnets affect each other.

f. Move your suspended magnet as it hangs in the water of an aquarium. Compare how the magnet behaves in the water with how it behaved in the air.

g. Obtain a galvanized pail and suspend your magnet in the center of the pail. Describe any differences in the behavior of the magnets in the pail and in the air in step a.

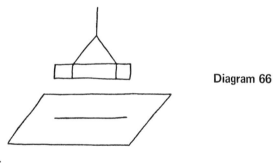

Diagram 66

VOCABULARY

    attract   repel

MATERIALS

    2 bar magnets   aquarium
    string   water
    paper   galvanized pail

PROCEDURAL INFORMATION

Be sure that the magnet is suspended far enough away from

iron objects as not to be affected by them. The magnets should respond the same in the water as in the air although the motion may be slowed slightly. The metallic frame corners of the aquarium may have some effect on the magnets. The effect of the surroundings of a magnet become more obvious when the students attempt to work within the confines of the pail. A discussion of how a compass can be used on board a steel ship may be generated by this activity.

EVALUATION

Shown a suspended magnet, describe the position you would expect another magnet would take if suspended in a different location in the same room.

Experience 3: CONSTRUCTING MAGNETS FROM DIFFERENT MATERIALS

OBJECTIVE

Demonstrate how magnets can be made.

Material	Before Stroking with Magnet	After Stroking with Magnet
toothpick		
straw		
sewing needle		
hairpin		
paper clip		
common pin		

Diagram 67

PROCEDURE

a. Bring one end of a toothpick near one end of a compass

needle and then the other. Observe if the toothpick attracts either end of the compass needle. Record your observations on the chart in Diagram 67.

b. Stroke the toothpick with a bar magnet by moving the magnet from one end to the other of the toothpick. Do not reverse the direction in which you move the magnet. Do this by raising the magnet several inches above the toothpick when moving it back to start the next stroke. Stroke the toothpick about thirty times.

c. Again bring one end of a toothpick near one end, then the other of the compass needle and observe if the compass needle is attracted to the toothpick. Record your observations in the 3rd column of the chart.

d. Repeat steps (a), (b) and (c) using the other items listed on the chart. You may add items to the chart. Record your observations for each item on the chart.

e. Circle the materials that affect the compass needle before stroking them with the bar magnet.

f. Circle the materials that affect the compass needle after they have been stroked with the magnet.

g. Compare the answers in (e) and (f). If an object affected the compass needle before it was magnetized, did it also affect the compass after it was magnetized?

h. How were the objects that affected the compass needle alike?

i. Describe any difference in how an object affected the compass needle, after it was magnetized, in comparison with before it was magnetized.

VOCABULARY

compass   magnetism

MATERIALS

compass	pin
bar magnet	paper clip
toothpick	hair pin
straw	sewing needle

PROCEDURAL INFORMATION

In doing this activity, the student should be encouraged to add as many items to his list as he would like to try. Things containing iron will affect the compass needle before being magnetized. However, after they are magnetized one end of the object will attract the needle while the other end repels it.

EVALUATION

Given a compass and a steel knitting needle, determine if the knitting needle is magnetized.

Experience 4: INFER THAT A COMPASS NEEDLE IS A MAGNET

OBJECTIVE

Observe a compass needle and a bar magnet and infer from similarities in their behavior that the compass needle is a magnet.

PROCEDURE

a. Observe the direction of sunrise and sunset. From this determine the approximate direction of north and south in your classroom.

b. Suspend a bar magnet from a string in the center of the room. When the magnet stops swinging, observe the directions in which the poles are pointing. Be sure to keep the bar magnet away from metal cabinets, files, chairs or desks.

c. Hold a compass a few feet from the magnet. Observe and record any similarities between the directions the magnet is pointing and the directions the compass needle is pointing.

d. Bring first the north pole, then the south pole of another bar magnet next to the north end of the suspended magnet. Repeat this using the end of the compass which is pointing north in place of the suspended magnet. Compare how the compass needle and suspended magnet respond to the bar magnet.

e. Using another compass, bring first the north-seeking pole, then the south-seeking pole near the south end of the magnet and then near the north-pointing end of the original compass. Compare how they behave.

f. Place a nail first at one end of the suspended magnet, then at the other end. Place the nail at one end of the compass needle, then the other. Is there any difference between the way the compass and magnet react to the nail compared with the second compass or magnet in steps d and e?

g. On the basis of your observations, would you infer that the compass needle is a magnet or not? Explain.

VOCABULARY

north-seeking pole   south-seeking pole
                non-magnetic

MATERIALS

2 magnetic compasses   string
2 bar magnets   nail, steel

PROCEDURAL INFORMATION

It is often difficult for a child to associate the magnetic characteristics with a magnet when it is with a non-magnetic object. For example, is it the nail which attracts the compass needle or vice versa? By comparing the reactions of known magnets such as a compass with other magnets and non-magnetic objects, a set of observations and inferences can be established.

EVALUATION

Given a magnetic compass, demonstrate how one can infer that the needle is a magnet.

Experience 5: DEMONSTRATING THE PRESENCE OF A MAGNETIC FIELD

OBJECTIVE

Observe the presence of a magnetic field and describe it.

PROCEDURE

a. Place a bar magnet on your desk top and cover it with a piece of transparent acetate. Carefully sprinkle a thin layer of iron filings on the acetate. Make a diagram of the results.

b. Describe any pattern which you can observe. Is there any observable concentration of iron filings?

c. Replace the iron filings in the container. Place the north end of one bar magnet about 2" from the south end of another bar magnet. Cover these with the sheet of acetate and again sprinkle iron filings over the parts of the magnets covered by acetate.

d. Make a diagram showing any patterns in the iron filings. Replace the filings in their container.

e. Place the north end of one bar magnet about 2" from the north end of another bar magnet. Cover these with a sheet of acetate and sprinkle iron filings over the acetate above the magnets.

f. Make a labeled diagram of your results. Replace the filings in the container.

g. Place the south ends of the two bar magnets about 2" from each other and cover them with acetate. Is there any difference between this and the results from the two north poles?

VOCABULARY

    magnetic field   field of force
                    lines of force

MATERIALS

    2 bar magnets   acetate
    iron filings   paper

PROCEDURAL INFORMATION

The iron filings line up with the lines of force which make up the magnetic field surrounding the magnet. The fields are concentrated at the poles and the reactions between like and unlike poles become very visible. The acetate permits the students to see the magnets yet prevents the iron filings from coming in direct contact with the magnets. This makes cleaning up afterwards a much easier task than removing the iron filings directly from the magnet.

EVALUATION

Given a horseshoe magnet, iron filings, and acetate, describe where the magnetic field would be and how it would appear if iron filings were sprinkled between the poles. Demonstrate that your description was correct.

Experience 6: IDENTIFYING THE RELATIONSHIP BETWEEN DISTANCE AND MAGNETIC FORCE

OBJECTIVE

Observe the effect of a magnet on different-sized objects at different distances from it and infer a relationship.

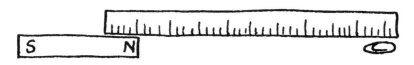

Diagram 68

PROCEDURE

a. Select several different objects which are attracted by a magnet and which have different masses. Make a list of the objects you select.

b. Place a bar magnet on a smooth surface next to a ruler as shown in Diagram 68. Place the first object from your list at the other end of the ruler. Is it attracted to the magnet?

c. Slowly slide the object towards the magnet being sure to keep track of the distance between the two. At some point, the magnet will attract the object. Record the distance between the two when the object is pulled toward the magnet. Repeat this three times and get a final average distance.

d. Repeat this procedure with each of the items on your list.

e. Was the distance the same for all objects? Explain.

f. From your results, what can you infer about the distance an object is from a magnet and the effect of the magnet?

g. What can you infer about the size of an object and the effect of the magnet?

VOCABULARY

mass

MATERIALS

metric ruler   bar magnet
several objects containing iron (paper clip, thumb tack, nails, needle, iron ore)

PROCEDURAL INFORMATION

The students should be encouraged to place a wide variety of objects on their list. The objects should vary in mass from a needle upwards to something about the size of the magnet. When the mass exceeds that of the magnet, the magnet will do the moving. Steel ball bearings of various sizes work very well for this activity since they roll so easily. A discussion of the influence of friction on the distance various objects move would be appropriate.

EVALUATION

Given three different objects containing iron and a bar magnet, order the objects on the basis of the distance which they must be placed from a magnet to respond to its attraction, and demonstrate the accuracy of the order.

10

How to Identify One's Location

The ability to identify the location or altitude of an object is a skill, the development of which relies heavily on mathematics. There are occasions when knowledge of the height of a tree or the width of a stream is desired. On other occasions, it may be important to find a building in an unknown part of the community, locate a community on a road map, or find a little-known community on a map of an unfamiliar country. Some elementary knowledge about angles, proportions, and coordinate systems will prove useful to the student facing any of these problems.

An elementary method for locating objects uses the clock. The object may be identified as being at 3 o'clock or 7 o'clock from a given object. From that, the use of the degrees in a circle is a next step. An object may be located at an altitude of 43° or 90° above the horizon or to the right or left of a given point. Using the compass to identify a position on the earth's surface by degrees is generally accompanied by the use of the cardinal points. As we attempt to become more precise in locating objects, various types of coordinate systems emerge. A person may be sitting in row 3, seat 5. A store may be located at 5th Avenue and 52nd Street. A crossroads may be found at 3-D on a road map. And, that little known foreign community may be at the intersection of 53° S and 35° E on a map.

The highly refined system of latitude and longitude is a universally-used coordinate system. Once a student becomes adept at using latitude and longitude, he is able to give the location of any site on earth. He is also able to find any location, given its coordinates. If he transfers this system to the celestial sphere, he has acquired the ability to apply a highly abstract, mathematically-oriented communications skill.

The accompanying experiences may provide the initial introduction for students in the use of the compass card for locating objects, or finding their altitude, and in the use of coordinate systems. Integrating these experiences with mathematics will increase the relevancy of the experiences.

Experience 1: CONSTRUCTING AND USING A COMPASS CARD

OBJECTIVE

Construct a compass card and measure the bearings of at least two prominent objects from one location.

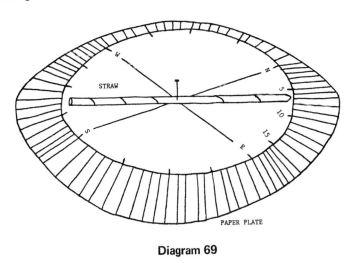

Diagram 69

Diagram 70

PROCEDURE

a. Turn a paper plate upside down and draw two lines which divide the plate into four equal sections as shown in Diagram 69. As you do this, count the ridges along the outside edge of the plate to be sure that there is an equal number of ridges between the lines.

b. Print the cardinal points of the compass (N,E,S,W,) in the appropriate positions. Starting with N as zero, count the ridges of the plate and number them clockwise by 5's.

c. Cut one end of a straw as shown in Diagram 70 to make a point. Push the point of a thumbtack up through the center of the plate from the underside. Place the straw on the tack so that its pointed end extends to the ridges on the edge of the plate and so that the straw can be rotated around on the plate.

d. Rotate the straw until the point is by the 10. The 10 is between north and east. If a surveyor were using your compass, he would record the reading as N10 ridges E. This would mean that the object being sighted bears 10 ridges to the east of north.

e. Locate north. Position the plate on a desk top so N is pointing north and each of the other directions is pointing the right way.

f. Select an object in the room and rotate the straw until it is pointing towards the object. Look through the straw and line the object up so it can be seen when sighting through the straw. Be sure not to move the plate while sighting on the object.

g. The point of the straw is close to one of the ridges on the edge of the plate. What is the reading for the object you have sighted? Record this on a data sheet.

h. Identify five other objects and take a bearing on them. Have another student take a reading of the same objects from the same location. Do your results agree? If not, ask your teacher for help.

i. A circle can be divided into 360°. The readings of compasses usually are divided into degrees. Cut out the circle with the degrees marked on it. Remove the straw and tape the circle to the paper plate, being sure that the center mark fits directly over the thumbtack. Replace the straw and again take readings of the same objects but using degrees this time. Again, compare your results with a classmate.

VOCABULARY

compass	direction
bearing	cardinal points
reference point	

MATERIALS

paper plate	straw
thumbtack	scissors
compass card ditto	
compass (if needed to determine north)	

PROCEDURAL INFORMATION

The ridges on the paper plate are convenient to use in taking a bearing on an object. It should be pointed out that degrees are move conventional and an internationally accepted standard. It should also be pointed out that the degree is a way of measuring the size of the angle formed between the line from the thumbtack and north, and the line which the straw makes as it points towards the object. Also, north has been used as the reference point for the activity for the sake of convention. It is possible for another point to be agreed upon by the class.

There are different types of compasses, each having variations in the way the directions or degrees are designated. The students may want to find the differences in a surveyor's compass, mariner's compass, and battleship compass. See Diagrams 71a, 71b, and 71c. They could then prepare a compass card for the compass of their choice that would fit their plate. When doing this, they should be sure that N on the compass card corresponds with the position of the original north on their plate. Also, the straw can be trimmed down so its point falls within the new circle. The directions call for the students to be supplied with a ditto of a compass card, divided into degrees, and the same size as the base of the paper plates they are using. The students can draw their own cards if preferred.

A discussion of the need for a compass card, possible causes for variations in the scales used, and the uses of a compass card

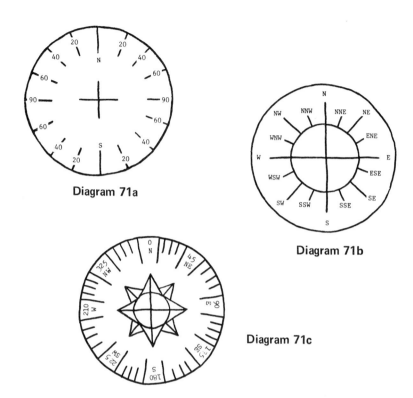

Diagram 71a

Diagram 71b

Diagram 71c

would be appropriate for the activity and would be helpful in introducing the next activity.

EVALUATION

Given a compass card, the reference point, and two objects, the student will give the bearings of the two objects using the compass cards.

Experience 2: MEASURING THE ALTITUDE OF OBJECTS

OBJECTIVE

Measure with a compass card the altitude of at least two objects from one location.

PROCEDURE

a. Mark the N-S line on your compass card "horizon."

b. Go outdoors and select an object such as a flagpole or building. To determine the altitude of the object, hold the compass card vertically so that the N-S line is parallel with the horizon. Sight through the straw to the highest point on the object and read the angle formed between the horizon and the straw. See Diagram 72. Record your reading.

c. Measure the distance from where you are standing to the base of your object. Record the distance.

d. On a large sheet of paper, draw a line to scale which represents the distance you measured. (Example: If the distance was 25 m., you may have a line 25 cm. long to represent the distance.)

e. Place your compass card along the line so that the thumbtack at one end of the line labeled "horizon" covers the distance line.

f. Rotate the straw as shown so it forms the same angle with the distance line as you measured outdoors. See Diagram 73. (It may be necessary to move your compass card and reposition the straw to the opposite end of the line before doing the next step.)

g. Make a mark on the paper at the end of the straw. Remove the compass card and place a ruler so you can draw a line which extends from where the thumbtack was, through the mark and on far enough so it passes above the other end of the distance line. See Diagram 74.

h. Draw a line perpendicular from the other end of the distance line until it touches the line drawn in g. This line represents the altitude of your object. Measure its length between the two lines and convert it to the same scale as you used to draw the distance line. Compare your results with the results of others using the same altitude.

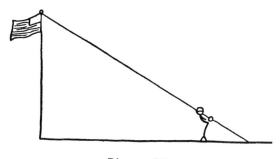

Diagram 72

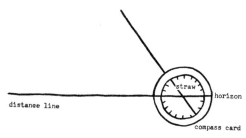

Diagram 73

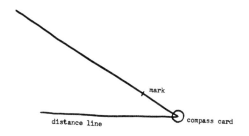

Diagram 74

VOCABULARY

altitude   scale
perpendicular   convert

MATERIALS

compass card   newsprint
ruler   tape measure (metric)

PROCEDURAL INFORMATION

The altitude of an object may be stated in degrees as in step b or in linear measurements as determined by doing the entire activity. If your purpose is to provide experiences in using a compass card, you may wish to use only steps a and b. If the altitude or height is to be determined in metric or English units, the remaining steps must be completed.

When selecting an object, it is desirable to select one such as a flagpole, tree, or side of a building that forms a perpendicular with the ground. This provides you with the important right angle at the far end of the base line. When marking in the angle measured with the compass card at the other end of the base line, the student is able to construct a right triangle. This right triangle is directly proportional to the one formed by the object, the baseline, and the line of sight. It has two angles and the baseline in common with the actual situation. The proportion between the baseline and the actual distance is the same between each of the other sides of the triangle in the student's diagram. Therefore, a direct measurement on the diagram can be used to calculate the actual lengths involved.

The student can use this process for finding the altitude of most objects. It is not necessary for him to know all the geometry involved in right angles to use the processes effectively. A few students may seek further information about the mathematics involved and should be encouraged in their seeking.

EVALUATION

Given a compass card, find the altitude of a given object.

Experience 3: IDENTIFYING LOCATION USING A SURVEYOR'S COMPASS

OBJECTIVE

Measure the bearing of an object from two positions and use this to determine the distance to the object.

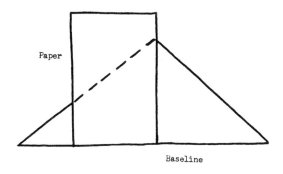

Baseline

Diagram 75

PROCEDURE

a. While standing in or near one corner of the back of the room, select an object in the front of the room for which you will determine the distance using the compass card.

b. Take a bearing on the object and record it.

c. Moving towards the other back corner of the room, measure off at least 5 m. from where you took the bearing. If space is available, move 10 m. Record the distance.

d. Take a bearing on the object from this new position and record it.

e. On a large sheet of paper, draw a line to represent the distance between the two locations from which you took the bearings. This line is the baseline and should be drawn to scale. If the original baseline was 10 m., you may scale it so 1 cm. represents 1 m. or .5 cm. represents .5 m. depending on the size of your paper.

f. Using your compass card or a protractor, mark off on the appropriate ends of the line the two angles you recorded. Draw a line from the end of one line through the corresponding mark and beyond it. Repeat this for the other end of the baseline and extend the line until it intersects the one from the opposite end of the baseline.

g. Slide a fresh sheet of paper along the baseline until one edge of it goes from the new angle to the baseline and forms a right angle where it comes in contact with the baseline. See Diagram 75.

h. Measure the length of the edge of the paper from the angle to the baseline. Using the same scale you used to draw the baseline, how far away from your baseline was the object you selected?

i. Measure the actual distance from the baseline to the object. If your answer is very different, ask your teacher to check your work.

j. Go outdoors and try to determine the distance to an object. Remember, the longer your baseline the more accurate your measurement can be.

VOCABULARY

baseline	protractor
bearing	intersect
scale	corresponding

MATERIALS

compass cards	large sheet of paper
meter stick	protractor

PROCEDURAL INFORMATION

The procedure used here is similar to that used in determining altitude in the previous experience. Rather than restricting the student to a right triangle, a piece of paper is introduced to form the right triangle. The paper provides an easy and relatively accurate way of constructing the perpendicular from the angle to the baseline. With a little direction, some of the students may be able to construct the perpendicular rather than using the paper.

It will be necessary to provide instruction on the use of a

protractor if the students have not had such previous experience.

It is conceivable that a student may want to determine the size of a pond, or the width of a lake or a river, in conjunction with a project on his environment. However, this same procedure can be used to estimate distances such as the length of a street, the distance to a building, the height of a tree, building, or flagpole, or even the distance to a landmark in a nearby city as viewed from a hilltop. A discussion of the advantages of indirect measurement might begin with the surveyor confronted with finding the width of a river and not being able to reach the other side. The procedure for determining the distance to the moon brings out the need for a very long baseline when sighting on a very distant object.

The students may want to invite a local surveyor into the classroom to find out additional information and add relevancy to the topic.

EVALUATION

Given a compass card and two positions, take a bearing on at least two objects from each position.

Experience 4: DEMONSTRATING A COORDINATE SYSTEM FOR DETERMINING LOCATION

OBJECTIVE

Demonstrate the use of a coordinate system for determining locations on a grid.

PROCEDURE

a. The location of the dot on the accompanying grid is 3N and 2E. Place a dot on the grid at each of the following locations (See Diagram 76):

2N and 4E   0N and 3E
4N and 2E   3N and 3E

Check your answers with your teacher's grid.

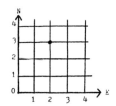

Diagram 76

b. Using the following grid, find these locations and place a dot on each (See Diagram 77):

60S and 45W   60S and 0W
15S and 75W   30S and 45W

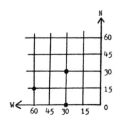

Diagram 77

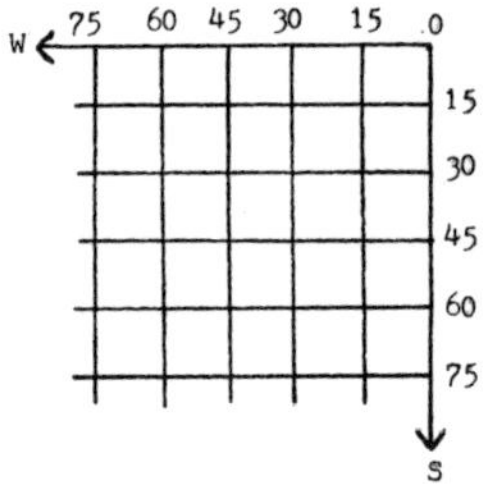

Again, check your answers with your teacher.

c. In step a, what direction was given first? Which direction was given first in step b?

d. Imagine that each dot on the following grids is a ship. Give the location of each ship shown. See Diagram 78a and 78b.

Diagram 78a

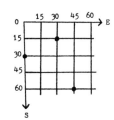

Diagram 78b

e. Locate the following on the accompanying grid and place a dot at the location. See Diagram 79.

45S and 15E   ON and 45E
30N and 30E   15S and 45E

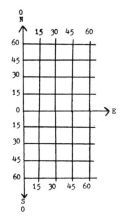

Diagram 79

f. On this grid each dot represents a city. Give the location of each city. See Diagram 80.

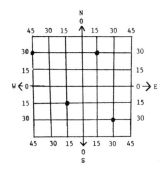

Diagram 80

g. The city shown on the following grid is located at 15N and 20E. Its location is given as 20E since it is about 1/3 of the way between 15E and 30E. See Diagram 81.

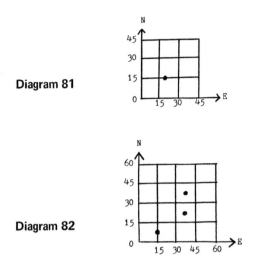

Diagram 81

Diagram 82

What are the locations for the above cities? See Diagram 82.

h. Examine some of the maps your teacher has put out for you. The lines on the map represent degrees. See if you can find the city which is at 39°N and 105°W.

VOCABULARY

degrees   grid

MATERIALS

variety of wall maps and atlases

PROCEDURAL INFORMATION

The teacher is able to monitor the progress of the student as he does each activity. If he encounters difficulty at any step, additional practice can be provided before the student proceeds.

Students can make a game out of any of the grids to provide practice or reinforcement. A variation of Tic-Tac-Toe can be played by pairs of students who identify the locations of their dots by calling off the correct coordinates. The first to get three or four dots in a row on the grid (horizontally, vertically, or diagonally) is the winner.

The terms latitude and longitude have not been introduced here although a teacher may wish to do so. Also, the maps selected for step i should be large and uncluttered, to assist in the transition from the grids to the maps.

EVALUATION

Given four locations on a coordinate system, identify the coordinates for three of the locations.

Experience 5: USING A COORDINATE SYSTEM ON A SPHERE

OBJECTIVE

Name and identify the latitude and longitude of a given position on a globe.

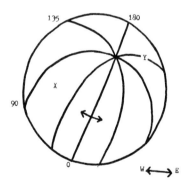

Diagram 83

PROCEDURE

a. Find the 0° line running N-S and label it the Prime Meridian. There are 360° in a circle. The diagram shows the earth divided into eight sections. How many degrees are there in each section? The number of degrees of longitude are numbered E and W of the prime meridian up to 180° which is half way around the earth from the prime meridian. Find the line opposite the prime meridian and label it 180°. The first lines E and W of the prime meridian should be labeled 45°. Label the rest of the lines. The prime meridian is the starting point for giving E and W locations. See Diagram 83.

b. City x is between____° and____°____of the prime meridian. City y is____°____ of the prime meridian.

c. The lines between the North and South Poles are meridians or lines of longitude. They help find locations of places___and ___ of the prime meridian.

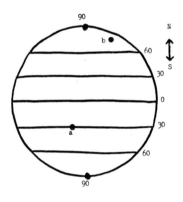

Diagram 84

d. On Diagram 84, label the 0° line running E and W on the equator. It divides the earth into a___and a__. It is the starting point for giving north and south locations.

e. City a is____°____ of the equator. City b is between ____° and____°____of the equator.

f. The lines going around the earth to the east and west are called parallels or lines of latitude. They help you find locations of places____and____of the equator.

g. Diagram 85 puts the two grids together to help give a more accurate location. City a is____°S of the equator and____° W of the prime meridian. City b is between 60° and 90°___of the equator and____° E of the ___. City x is ___°N and between

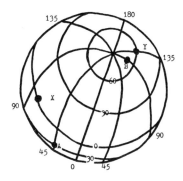

Diagram 85

___° and ___° W of the prime meridian. City y is ___° N of the ___and ___°___of the prime meridian. Check your answers with your teacher. If they are correct go on to the next step.

h. The grid you have been using is like the one found on an earth globe. Use a globe and give the locations of the following cities.

New York City	___° N ___° W
London	___° N ___° W
Melbourne	38° ___145° ___
Moscow	___° N ___° E
Los Angeles	34° ___118° ___
Rio de Janeiro	___° S ___° W

VOCABULARY

degree	longitude
equator	meridian
latitude	parallel
prime meridian	

MATERIALS

globe

PROCEDURAL INFORMATION

In Experience 4, the student was involved in activities using a coordinate system for giving locations. The transfer from the two-dimensional grid to the representation of the three-dimensional globe will require additional practice for some students. It will prove helpful for others to have access to the globe, early in the activity, to enable them to visualize the situation.

It is customary to give the N-S direction first. In doing so, the student will have fewer problems if the direction is given as "degrees north of the equator" or "degrees south of the equator." In the same manner, an E-W direction should be accompanied by E or W "of the prime meridian." Latitude and longitude can be very confusing to students if they are not introduced to the concepts slowly.

EVALUATION

Given the latitude and longitude of five cities, identify four of the cities correctly.

11

How to Measure Time

Man's need to be able to increase the precision with which he measures time appears to be directly related to his progress. There was a time when being able to tell the time or measure a span of time by the relative position of the sun or stars was adequate. As life became more complex, the need for greater accuracy was accompanied by improvement in our clocks and other timing devices. It is now possible to measure minute intervals of time with almost unbelievable accuracy.

The relativity of time is a difficult concept for children. "A short time" and "a long time" may be used to describe anything from a minute to years. Add to this that the same person, on different occasions, uses the term "a short time" to describe an event that lasted five minutes or one that lasted five months. Then add to this the arbitrary methods that have been devised for measuring time, establishing time zones, or determining daylight savings time, and the complexity of the task of telling time by children becomes apparent.

The following experiences are an effort to provide a variety of time-determining situations. It is hoped that the variety will provide a meaningful way of looking at time for some students.

Experience 1: DEMONSTRATING THE RELATIVENESS OF TIME

OBJECTIVE

Observe the similarities or differences between people's age and how they measure time in terms of past events.

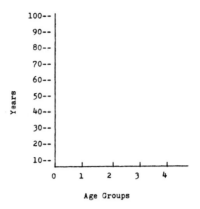

Diagram 86

PROCEDURE

a. Describe what you think "a long time ago" is.

b. Conduct a survey of the following people:

    1. yourself

    2. someone younger than yourself

    3. someone about the age of your parents

    4. someone about the age of your grandparents

Ask them to think of something that happened in their lives as far back as they can remember and to estimate how long ago this event happened.

Record their answers.

    1. __ years ago

    2. __ years ago

    3. __ years ago

    4. __ years ago

c. Make a chart of the answers from all of the students in the class by age groups of those questioned.

d. Find the average number of years ago for each group by dividing the total number of years by the number of people questioned in the age group. See Diagram 86.

e. How long ago was "a long time ago" for the oldest group?

How long ago was "a long time ago" for the youngest group?

How do you explain your results?

f. List some ways people measure time.

g. How come we tell time in hours?

h. How far back do you think time goes?

VOCABULARY

average   incident   time

MATERIALS

graph paper   pencil

PROCEDURAL INFORMATION

The students may wish to survey several persons and graph their own results as well as contributing to the class effort. Their results may lead them to the hypothesis that "a long time" is shorter for young people and longer for older people.

The request for a list of ways people measure time is very open. Responses may vary from mechanical devices to natural events to time intervals. Some items listed might include: watch, clock, radio, telephone, factory whistle, minute, hour, day, month, year, sun, sunset, sunrise, moon, stars, earth's rotation, or earth's revolution. The students can classify their responses as to whether they were natural events or man-made. A discussion of the definition of man-made units of time listed such as "how long is an hour?" would be a good introduction for some of the following experiences.

EVALUATION

Given the following statements, predict if the person saying it was younger, the same age, older, or much older than you are, and explain your answer.

1. We were the first people on our street to have a radio.

2. It wasn't too long ago that everyone walked to school.

3. We just had an ice cream cone.

Experience 2: IDENTIFYING TIME ZONES AND THEIR FUNCTION

OBJECTIVE

Distinguish between a.m., m., and p.m., and identify time zones and their functions.

PROCEDURE

a. It appears that the sun circles the earth once a day. There are 360° in a circle, so if there are 24 hours in a day, how many degrees does the sun appear to move every hour?

b. The starting point for telling time is the 0° line of longitude which is called the _____. What large city is located close to this meridian?

c. Every 15° of longitude is a time meridian which is in the center of a time zone. The time zone extends 7-1/2° east and 7-1/2° west of a meridian. There are 24 time zones around the world.

d. With other members of the class, arrange 24 chairs in a large circle. Point out the position of the sun and label the chair directly in front of it NOON. Label the other chairs, moving to the left from Noon, marking the first chair 11:00 a.m., the second 10:00 a.m., and on around to Noon again.

e. Have people take a seat in the 24-hour clock. Each one should determine the meridian for his position and make a sign showing it.

f. After determining the time and meridian for your seat, use a map and identify either a city or geographic feature located on or near your meridian. Add this information to your sign. When everyone has all of the information on their signs, return to your seat in the circle.

g. What time is it in London according to your clock? Compare the times for other locations. Discuss what people would be doing at your location for that time.

h. The earth rotates counterclockwise 15° every hour. Imagine that one hour has passed and change chairs and show this. Take your sign with you but notice the time label on the chair you move to.

i. Describe which way the sun appeared to move in relationship to you.

j. Now which meridian is the sun directly over? If m stands

for meridian, why is noon called 12m?

k. Imagine that five hours pass and slowly change seats, one hour at a time, to show this change. Observe the apparent motion of the sun. These locations are now after noon (12 m) or at hours labeled ____. The locations before noon (12m) are at hours which should be labeled ____ times.

l. Continue changing positions until 24 hours have passed. Where is London now?

VOCABULARY

a.m.	standard time
p.m.	meridian
m	longitude
time zone	prime meridian

MATERIALS

atlas		paper for signs
	24 chairs	

PROCEDURAL INFORMATION

If 24 hours or the space for 24 chairs is not available, the chairs may be spaced at intervals of 2 or 3. It is desirable to involve as many students directly in the activity as possible. Some of the locations that can be used might include:

London	0° longitude
Moscow	45° E
Rhutan	90° E
Seoul	135° E
Fiji Islands	180° (International Dateline)
Anchorage	135° W
Chicago	90° W
Rio de Janeiro	45° W

If no city is close by, geographic features may be used such as islands, bodies of water, or mountains.

It is possible to demonstrate daylight saving time by repeating the circle procedure. In this case, when the sun was directly on the prime meridian, it would be 1:00 p.m. rather than 12 m. This could be accomplished by moving all of the chair labels ahead one hour. If the students are successful with this, you may wish to assign dates to the chairs and observe what happens at the International Dateline.

EVALUATION

Given a diagram of the 24-hour clock made by the class, without labels on the chairs, and the relative position of the sun, tell what time it is at any three meridians.

Experience 3: CONSTRUCTING A TIME LINE FROM TREE RINGS

OBJECTIVE

Identify the age of a tree by its rings and construct a master chart from growth rings identified in a tree's cross section.

PROCEDURE

a. Examine a cross section from a tree. Are all of the rings the same width?

b. If each ring represents one year of growth for the tree, what would have caused it to grow more some years than others?

c. Place a piece of paper on the cross section so it reaches from the bark to the center of the section. Mark each ring on the paper. Starting with the layer just under the bark, write the date of the current year next to it. Counting backwards one year for each ring, how old is the section? Label the center ring with the year it was planted. Label the rings on the section which grew during special years such as the year you were born, the year you entered school, and the year the first man reached the moon.

d. Examine the diagram of the cross section cut in 1970 (Diagram 87). Make a long strip of paper and lay it down so the bark from this diagram is at one end of the paper. Mark off each tree ring on the paper. Label the center ring 1970 since that was the year the tree was cut. Count off and label each fifth ring by year.

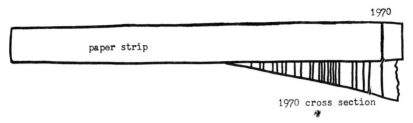

1970

paper strip

1970 cross section

Diagram 87

e. Place the cross section from the tree cut in 1952 on the paper strip so the outer ring is next to the ring for 1952 from the 1970 cross section. Mark off the additional rings from this section and date them. Do the same thing for the cross sections cut in 1935 and 1913.

f. Using the master chart of the tree rings which you just made, how far back in history do your tree rings go? In what year was there the greatest amount of rainfall?

g. What can you say about the rainfall from 1912-1918? From 1941-1946?

h. Identify famous events in history and label them on the master chart by the appropriate ring.

i. What difference would it make on your master chart if the cross section came from the tree trunk or a high branch?

VOCABULARY

    cross section   dendrochronology

MATERIALS

    tree cross sections   cross-section diagrams

PROCEDURAL INFORMATION

Diagrams 88a, 88b, 88c, and 88d of the cross sections were devised for this experience and can be used as a guide to make separate diagrams for each student. It would be possible for students to bring in cross sections and construct their own master chart. If this is done, there are a few precautions which should be kept in mind.

This tree was cut in 1970. The outer ring is the growth for 1970.

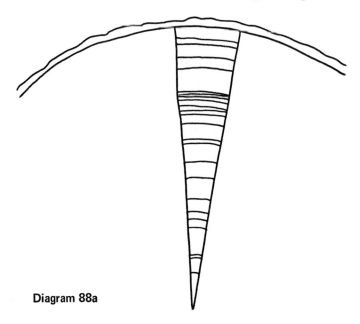

Diagram 88a

This cross section is one taken from a tree cut in 1952 for construction of a bridge.

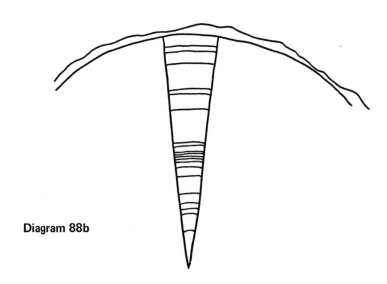

Diagram 88b

This cross section was taken from a log cut for a telephone pole erected in 1935.

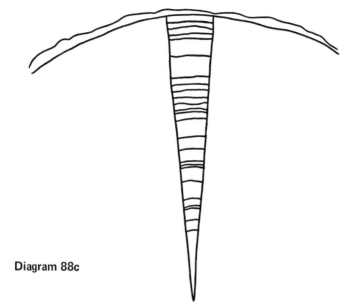

Diagram 88c

This cross section was obtained from a log cabin built in 1913.

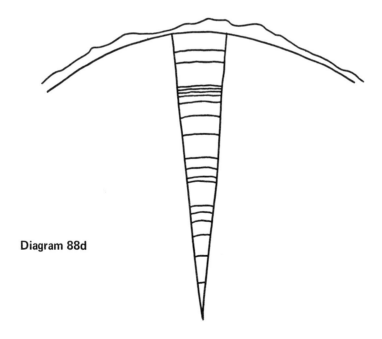

Diagram 88d

Some rings vary in thickness being thin on one side and fat on the other. This may be the result of one side being shaded, touching something, or being exposed to strong winds or cooler temperatures. If this is the case, most of the thin rings will be on one side. Since you are concerned with the general pattern of the rings rather than their actual thickness, you would select a section between the extremes.

Trees grown in the same year in different parts of the country will have different ring patterns depending on growing conditions. A master chart would have to represent trees from the same general location, if it was to have any validity. It would, also, be desirable to use trees with at least 25 rings, which were cut in known years. After the master chart is established, trees with unknown cutting dates from the same area could be dated.

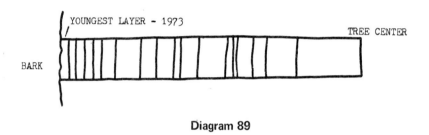

YOUNGEST LAYER - 1973

TREE CENTER

BARK

Diagram 89

EVALUATION

Given a fresh tree stump with the accompanying pattern, tell whether you were alive when the tree was and which is older, you or the tree. See Diagram 89.

Experience 4: IDENTIFYING A SPLIT SECOND

OBJECTIVE

Observe a high-speed photograph and infer how long a split second might be.

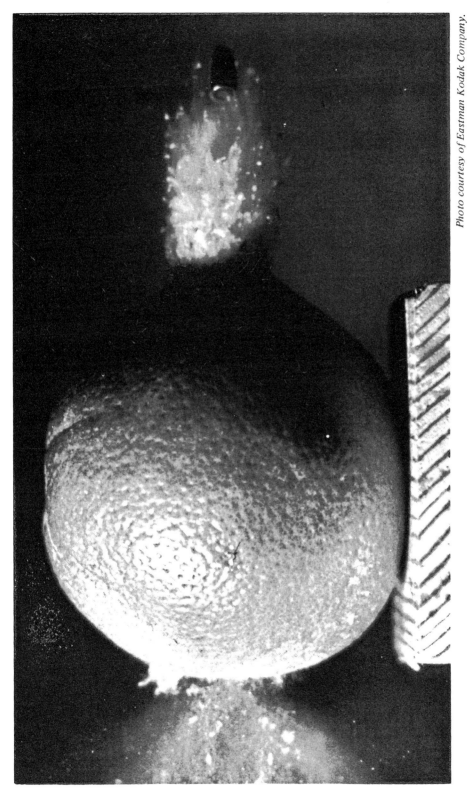

Diagram 90: **Freezing a Bullet**

187

PROCEDURE

a. Observe the photograph entitled "Freezing a Bullet" and record your observations. See Diagram 90.

b. What would you infer about what was happening in the picture? Would you say this happened in a split second?

c. One often thinks that a second is a short interval of time and very little is learned or changes in short time intervals. The picture shows a .22 caliber bullet, traveling at Mach 1 (1100 ft./sec.). The flash lasted 1-millionth of a second. How many millionths of a second are there in one second?

d. There are __ seconds in a minute and _____ seconds in one hour. There are __ hours in a day or _____ seconds in one day. 1,000,000 seconds are equal to about __ days Therefore, one second is (bigger, smaller, about the same) when compared to 11 days, as 1/1,000,000 of a second is when compared with one second.

e. How wide is the orange in the picture?

f. How long is the bullet?

g. How far is the bullet from the orange?

h. The bullet was traveling at 1100 ft./sec. What was its speed in inches/sec.? (100 ft. x 12 = _____)

i. How long a time has it been since the bullet left the orange? distance from orange
  inches/second

j. How much time did it take the bullet to go through the orange? width of orange
  inches/second

k. What are some reasons why high-speed photographs such as this are useful to scientists and engineers?

VOCABULARY

high speed photograph   Mach 1
millionth of a second   second

MATERIALS

"Freezing a Bullet" photograph (Xerox copies of this photograph) or a similar high speed photo
ruler

PROCEDURAL INFORMATION

A split second is often used to describe the time it takes for a very rapid incident to occur. Rarely have students attempted to identify what happens in a split second. The time it takes the bullet to pass through the orange may well have been inferred to be a split second. In this case, they are able to calculate the length of their split second. The mathematical assistance given will vary with the students and is secondary to the recognition of what can occur in a split second.

Other high-speed photographs showing some action can be used for this experience. Photographic anthologies would be a good source for additional pictures and would provide the speed of the projectile. The pictures might show a baseball player swinging a bat, a bowler, a golfer swinging, a racing car, or a motorcycle.

EVALUATION

Describe what is me nt by the term "a split second."

Experience 5: MEASURING TIME BY THE SUN

OBJECTIVE

Demonstrate the relationship between the position of the sun in the sky and the time of day and how this relationship can be used to tell time.

PROCEDURE

a. What is the total number of hours in one day? How many times does the hour hand of a clock go around the dial for one day?

b. Where is the sun at noon?

c. Place a large sheet of paper on the floor and draw a 24-hour clock on it as shown in Diagram 91. Use a bare light bulb in a lamp to represent the sun. Position it 10-15 feet from

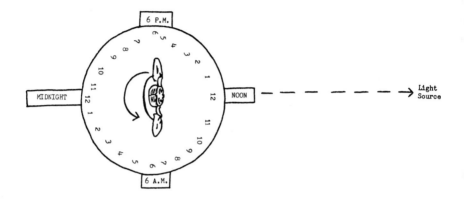

Diagram 91

the 24-hour clock on a line directly out from 12 o'clock noon.

d. Stand in the middle of the clock facing the sun. Hold your open hands, palms forward, along side your eyes so as to make a straight line between 6 p.m. and 6 a.m.

e. The earth rotates counterclockwise from west to east so if you represent the earth, you must turn slowly to the left. In what direction does the sun appear to move, as you move to the left?

f. When you are facing 6 p.m., in what position does the sun appear to be? What happens to the sun as it gets later, as you continue to revolve?

g. Where is the sun at midnight? At what time does the sun appear to rise?

h. If you continue turning, what time is it when you are back directly facing the sun?

VOCABULARY

    rotate   sunrise   sunset

MATERIALS

light bulb and socket
large sheet of paper
marking pen
string (to make circle for clock face)

PROCEDURAL INFORMATION

It is highly unlikely that most students will ever have to rely on telling time solely on the sun's location. However, it does help to establish the basic relationship between the earth-sun positions and our system of telling time. Once the students have established the relative positions of the sun for morning, noon and afternoon, they may wish to verify their information. This can be done by moving the paper outdoors. It should be placed so that at 12 noon the sun is in the same relative position to the paper as it was inside.

When outside it may become apparent that the sun is not directly overhead at noon. Actually, the sun is directly over one's meridian, or at its highest point in the sky, at noon, solar time.

It should be brought out in a discussion that the sun's motion is an apparent motion.

Diagram 92

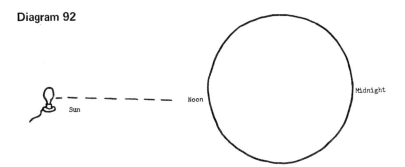

EVALUATION

Given a clock face showing only midnight and noon, and a lamp placed 15 feet from the noon position of the clock face, state a time and position the earth (yourself) appropriately for that time in relationship with the sun. Repeat this process, naming at least three different times, and assuming the correct corresponding position at least twice. See Diagram 92.

Experience 6: CONSTRUCTING A TIMING DEVICE

OBJECTIVE

Construct a device for measuring time intervals.

Diagram 93

PROCEDURE

a. Obtain two glass jars of the same size and remove the covers from both of them. Make a hole with a small nail, about 3 mm. in diameter in the center of one cover. Set the other cover aside for later.

b. Fill one jar with dry sand and place the cover with the hole on it.

c. Place the mouth of the other jar on top of the cover on the first jar. Using masking tape, tape the two jars securely together as shown in Diagram 93.

d. Estimate how long it will take the sand to flow into the empty jar if you invert the jars. Check your estimate with a watch having a second hand.

e. When the sand is in the bottom jar, invert the jars again but this time mark the level of the sand every 30 seconds. Use a wax pencil which will mark on glass.

f. Use your sand clock to time different events such as the time it takes someone to do 10 situps.

g. Take the other jar top and make a very tiny hole in the center of it with a thumb tack. After you have finished using the sand clock, remove the sand and replace it with water. Fasten the two jars together and calibrate your water clock. Use it to time some different events.

h. Which of the clocks did you like best? Explain.

i. If you were to exchange the tops from the two clocks, what changes would you expect in your clocks?

VOCABULARY

calibrate   minutes   seconds

MATERIALS

two similar jars with screw top covers
watch with second hand
small nail   sand
thumb tack   masking tape
wax pencil   water

PROCEDURAL INFORMATION

When calibrating their clocks, the students should observe when the steady flow of sand or water stops. This should be the end of the time the clock can measure.

Some research will reveal the early uses of sand and water clocks. Students may want to bring in samples of egg timers and other timing devices. By timing the egg timers they can find the time interval each one measures. Discuss the idea of timers that time given functions, compared with the function of modern clocks.

Given a plastic bag, a watch, and a styrofoam cup of salt, make a device that will measure 2-minute time intervals.

Experience 7: MEASURING GEOLOGIC TIME

OBJECTIVE

Identify and describe at least one way by which scientists measure the age of the earth.

PROCEDURE

a. Some minerals contain elements which change from being radioactive to non-radioactive elements. When this happens the change takes place at a definite rate of time. Since nothing changes this rate, it can be used to determine the age of the mineral. The rate at which a radioactive material becomes a non-radioactive material is called half-life. Each radioactive element has its own half-life as shown. If a scientist examines a mineral containing uranium-238 and can tell that half the uranium has changed to lead, he knows the mineral would be 4,500 million years old.

Mineral	Half-Life in Years
uranium-238 to lead 206	4,500 million
potassium-40 to argon	1½ million
rubidium-87 to strontium-87	60 million
carbon-14 to carbon-12	6 thousand
thorium-232 to lead	14 million

b. Meteorites have been found that contained half uranium-238 (U-238) and half lead-206. How old must the meteorite be?

c. Remains of early man have been found containing minerals with about half the thorium-232 changed to lead. About how long ago must this man have lived?

d. Some of the earliest grass fossils were found with minerals containing 10 grams of strontium-87. How old must the grass have been?

e. On the basis of rock structure, radioactive materials, and fossil formation, scientists have divided the history of the earth into sections called eras and subsections called periods. Using resource materials complete the information on the chart in Diagram 94.

f. If you found a fossil of a fish, how do you think you could estimate its age?

g. If you saw nine different layers of rocks in a cut along a highway construction site, which one would you expect to be the oldest? If fossils were found in some of the layers, where would you expect to find the oldest ones?

Era	Million of Years Ago	Period	Evidence of Life Types
Cenozoic	Present-63	Cenozoic	
Mesozoic	63-230	Cretaceous Jurassic Triassic	
Paleozoic	230-600	Permian Pennsylvanian Mississippian Devonian Siberian Ordovician Cambrian	
Precambrian			

Diagram 94

VOCABULARY

radioactive   era
non-radioactive   period

element	fossils
half-life	prehistoric
billion	meteorite

MATERIALS

resource books and visuals of the geologic time scale

PROCEDURAL INFORMATION

The question of how radioactive materials got inside early man or grasses should be anticipated. A discussion of radioactive materials in today's environment and their effect should be based on factual information rather than hearsay. The students can gather such information for examination from library sources or the local civil defense unit.

Radioactive dating has made it possible for scientists to be more accurate in their determinations of the age of different plant and animal remains. Geologic timetables are based on evidence of plant and animal life found in fossils and diggings and have been used for approximate dating of specimens. The accuracy of many of these approximate dates has been refined through the use of carbon-datings and other radioactive dating techniques.

EVALUATION

Select one way which scientists use to determine the age of prehistoric remains and briefly describe how they would date them.

12

Inquiries into the Motion of Moving Objects

A child can look at the sky and be very aware of the fact that there are objects in it which move. He realizes that the sun is in different places at different times. He knows that the moon is visible at nighttime but that there are times when it is also visible during the day.

The knowledge that the clouds, earth, moon, sun, and many other objects move is not enough. The questions which follow deal with how do they move and what are the effects of these motions. Efforts to inquire into these motions with children require abstract thinking. Since the children are unable to stand and watch some of these motions or to move to another frame of reference to observe them, models and long-term observations are used in an effort to create understanding.

Many of the experiences which follow involve the students in identifying frames of references and observing motion within these frameworks. Collecting data on their observations and developing models based on this data, provides the child with his own basis for understanding how these objects move and the effects of these motions.

Experience 1: OBSERVING PATHS OF MOVING OBJECTS

OBJECTIVE

Observe and describe the different paths which some moving objects follow.

PROCEDURE

a. Sprinkle a layer of flour onto a sheet of black construction paper. Place the paper, floured side up, so one edge is against the wall. Roll a marble at any angle towards the wall so that it rolls over the construction paper. Observe the trail in the powder left by the rolling marble and complete the diagram showing its path after striking the wall. See Diagram 95.

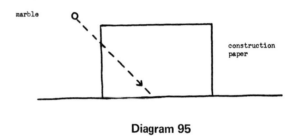

Diagram 95

b. Describe the path made by the marble before striking the wall.

c. Describe the path of the marble after striking the wall.

d. The path of the marble before it struck the wall and the path after it struck the wall form a(n) _____ at the wall where the direction changed.

e. Name some moving objects which move in a straight line or path. Identify any moving objects you can think of which follow an angular path.

f. Cover a piece of cardboard 12" x 24" with black construction paper. Sprinkle a layer of flour over the construction paper and place the paper on the floor with one end resting on a pile of books.

g. Roll a marble at an angle with some force so it will roll partly up the incline and down again.

h. Observe the pattern left by the marble and complete the diagram to show its path. See Diagram 96.

i. Roll the marble again when only one book is used to tilt the paper. Describe the differences in the two paths of the marble.

j. Place five books under one end of the paper and roll the marble again. Compare this path with the two previous paths. What name is given to a path shaped like these? Name some examples of objects which travel in paths shaped like these.

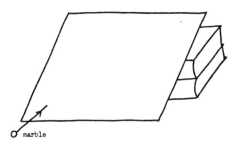

marble

Diagram 96

k. Fasten a marble to the top edge of an old record with some modeling clay.

l. Place the record on a phonograph and observe the path of the marble. Complete the diagram to show the marble's path. See Diagram 97.

m. Describe the path that the marble traveled.

n. What path do you think the marble would follow if the clay was not used? Show this on your diagram and list some other objects that travel in circular paths.

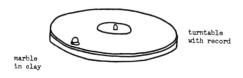

turntable with record

marble in clay

Diagram 97

o. Tie together the two ends of a piece of string which is 20 cm. long.

p. Place a piece of construction paper on a large piece of cardboard and draw a straight line about 8 cm. long across the center of the paper. Place a thumbtack firmly at each end of the line. Loop the string around the two thumbtacks and place a pencil inside the loop so it stretches the string as shown in Diagram 98. Move the pencil around the tacks keeping the string tight as you draw a line.

q. Describe the shape of the line made by the path of the pencil. Draw it on the diagram. Are there any objects you know of that travel in paths shaped like these?

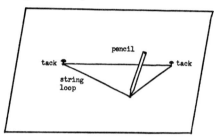

Diagram 98

r. Place a pencil or dowel in the center of a wagon wheel so you can hold onto it as it rolls. Tape a piece of chalk firmly to one edge of the wheel. Be sure that the chalk is long enough so it will mark on the chalkboard as the wheel is rolled in the chalk tray.

s. Roll the wheel along the chalk tray and observe the path of the chalk as it revolves around the center of the wheel. Describe the chalk's path and show it on the Diagram 99.

t. Are there any examples of an object revolving around another moving object which you can think of? List them.

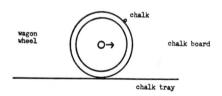

Diagram 99

VOCABULARY

angular   parabola
curve   ellipse
              revolve

MATERIALS

flour   clay
marble   phonograph
large cardboard   thumbtacks
string   paper
chalk   tape
wheel, 12 to 25 cm. in diameter (from wagon or trycycle)
old phonograph record

PROCEDURAL INFORMATION

The paths different objects follow present such variety as a straight line, angular, circular, parabolic, elliptical, and others. An examination or research of the paths of different moving objects may be necessary as students attempt to place different objects on their lists. These lists may include anything from the path of a golf ball to the path of a comet.

When rolling the marble up the inclined paper to form a parabola, the marble must be rolled so it approaches the paper at an angle. If being rolled from the left side of the paper, it should approach the left corner with enough force to carry it about halfway up the paper before it reaches the peak of its curve and rolls back. The students may want to compare the curves formed when the marble is rolled at different angles or with different forces.

The first paths the student examines are the straight path, such as that of a ball rolling on a smooth floor, and the angular path resulting from an obstacle changing the direction of a moving object, as in the case of colliding billiard balls. The marble moving up the incline traces a parabola which can be compared with the paths of projectiles such as baseballs, arrows and bullets. The revolving marble's path is circular and can be compared with that of the earth's orbit, which is nearly circular. Comets are the best example of celestial objects which have elliptical orbits, such as those produced by the string and thumbtacks.

The path obtained in step s would be similar to that of the moon as it travels around the earth. The axis of the wheel could be compared with the earth moving in its orbit and the chalk would represent the moon in its orbit. If a chalk tray and board are not available, fasten newsprint the height of the wheel on the wall and roll the wheel on the floor.

EVALUATION

Different objects travel in different paths. Identify four objects which travel in different paths and describe their paths.

Experience 2: MEASURING THE MOTION OF CLOUDS

OBJECTIVE

Measure and compare relative motion of clouds.

PROCEDURE

a. Go outdoors and select a location where you can get a clear view of the sky. Place your mirror on the ground and check your view of the sky in the mirror.

b. Looking in the mirror, watch a cloud move across it.

c. Position the mirror so that its longest sides correspond with the direction in which the clouds are moving. Place a ruler along one of these sides with the 0 end at the end of the mirror where the clouds first appear.

d. Select a cloud as it first appears on the mirror and measure how far it travels on the mirror in 10 seconds. Record your results.

e. Compare your results with the results of other students in the class.

f. Go outdoors and repeat this activity on at least five other cloudy days. When you have completed this, show your results on a graph where the distance the clouds traveled is on the y axis and the days measured is on the x axis.

g. Can you determine the actual speed of the clouds from your graph? What do the graphs show?

VOCABULARY

relative speed   rate

speed

MATERIALS

watch with second hand or stopwatch
graph paper   ruler

mirror

PROCEDURAL INFORMATION

There are times when one capitalizes on children's likes to provide motivation for achieving an objective. In a sense, this experience does that. Many children enjoy watching the clouds. This experience provides them with the opportunity to get outdoors and watch the clouds move for the purpose of collecting data for their graphs. However, several valid learnings can come from this besides constructing a graph. The students develoṕ a standard of their own for making indirect comparisons of the rate of motion of clouds. In doing this activity, it may become apparent that some children relate the movement of the clouds to the rotation of the earth. Misconceptions such as this can be clarified by comparing the rate of rotation with the speed of the wind.

This activity is one which requires a partially cloudy sky. An abundance of small fast-moving clouds would be ideal. If the ideal is not possible, the time span may be altered or the distance, rather than the time, could become the constant. In either case, a discussion of the comparison of relative cloud speed, rather than actual speed, is important.

EVALUATION

Given a graph showing the relative motion of clouds for

several different days, identify the days the clouds were traveling fastest and slowest and how much faster they were going on the fastest day compared with the slowest day.

Experience 3: INFER THE RELATIVE MOTIONS OF THE EARTH AND SUN

OBJECTIVE

Observe the effects of changes in the relative position of the sun and infer the relative motions of the earth and sun.

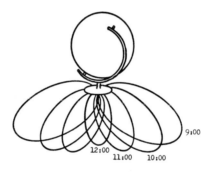

Diagram 100

PROCEDURE

a. Very early in the day, place a globe on a flat surface such as a sidewalk or parking lot. It should be positioned so it will be in the sunshine all day and so that its axis points toward North. Use a compass to determine North.

b. Trace around the base of the globe with a piece of chalk and place a mark on the surface to indicate the position of the axis, in case the globe must be moved during the day.

c. Trace around the shadow of the globe and mark the time you did this on the shadow.

d. Return to the globe each hour to trace its shadow and record the time.

e. How do you explain the change in the position of the globe's shadow?

f. On Diagram 100, draw in where you think the sun would have been to produce each of the shadows. From this diagram, one might infer that the ___ moved and the _____ remained still. Does the sun revolve around the earth?

g. Astronauts have indicated that they have observed the earth rotating. Is it possible that the same shadows could have been produced by the earth rotating and the sun remaining still?

h. Place one end of the dental cotton swab on a piece of double stick tape and cut the tape around the cotton. Locate your community on the globe and tape the cotton swab upright on it.

i. When the sun is high in the sky, slowly rotate the globe counterclockwise through a day while the sun remains in the same position. Observe the shadows made by the cotton and compare them with the shadows made by the globe.

j. How do you explain the change in the position and size of these shadows?

k. From what you have done and what you know about the earth and sun, what do you infer about the motion of the earth and sun?

VOCABULARY

relative position	rotation
shadow	revolution

MATERIALS

globe	compass
chalk	cotton
double stick tape	

PROCEDURAL INFORMATION

The students are involved in observing the changes in shadow size and position. From their observations, they should be able to make inferences about the rotation and revolution of the earth and the relatively fixed position of the sun.

Since the motions of the earth and sun cannot be measured directly, we must rely on indirect methods for such measurements. This results in measuring the relative motions of the two bodies. These relative motions, plus information gained from other sources such as astronauts and the Foucault Pendulum, support the hypothesis that the earth both rotates and revolves while the sun appears to remain stationary in relation to the earth.

Several discussion topics and extensions of this experience may be introduced either by the teacher or students. Included in these would be the variations in the length of day and night, the cause of day and night, the relative positions of the sun and earth in the different seasons, the seasonal variations which would occur in the length of shadows, or the reliability of shadow length for telling time.

EVALUATION

Given a stick positioned in the ground, record the position and size of its shadow at one hour intervals, for at least three hours, and state any inferences about the relative movement of the sun and earth that can be made from these observations.

Experience 4: OBSERVING THE MOON'S MOTION

OBJECTIVE

Observe and identify the moon's apparent and real motion around the earth.

PROCEDURE

a. Find out when the moon is visible during the daytime.

b. Select a day when the moon is visible for several hours and observe its position at one hour intervals. Record your observations on a diagram in which you show the position of the moon relative to a building or other stationary object.

Indicate the time each observation was made next to each moon you draw on the diagram.

   c. In what direction does the moon appear to be moving from hour to hour? Show this on your diagram.

   d. Select a series of three to five days when the moon will be visible at the same time. Observe the moon at the same time on each of these days and record your observations on a diagram. In your diagram, show the position of the moon in relationship to a building or other stationary object and record the date of your observation.

   e. In what direction does the moon appear to be moving from day to day? Indicate this on your diagram.

   f. One of your diagrams shows the real motion of the moon in terms of the earth's motion and the other shows the apparent motion of the moon. Examine your diagrams and discuss them with other students. Which one do you think shows the real motion of the moon? Explain.

VOCABULARY

   apparent motion   real motion
                  relative motion

MATERIALS

   drawing paper   crayons
   reference tables of moon's motion

PROCEDURAL INFORMATION

   In a departmentalized situation, it may be desirable for the students from several classes to share their observations in order to obtain the necessary data for this experience.

   The moon is visible during the day from its new crescent to gibbous, and old gibbous to crescent, phases. Observations at hourly intervals will reveal that the moon appears to be moving from east to west in the northern hemisphere or right to left on the diagram. Observations made at the same time on a series of different days will reveal a west-to-east motion or left to right on the diagram. See Diagram 101.

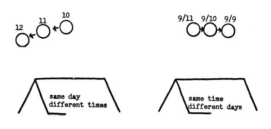

Diagram 101

The very rapid west-to-east rotation of the earth on its axis causes the moon to appear to be moving from east to west when viewed hourly. However, its position in regard to the sun is relatively unchanged. When viewed at the same time on consecutive days, the motion of the moon places it in a different position in the sky in relation to the earth and sun. Therefore, the hourly motion is an apparent motion while the daily change represents the real motion of the moon.

EVALUATION

Given a diagram of the moon and earth, identify with an arrow the real motion of the moon.

Experience 5: MEASURING THE SUN'S APPARENT MOTION

OBJECTIVE

Measure and predict the apparent movement of the sun by plotting shadows on the compass disc.

PROCEDURE

a. Fasten the compass card to the cardboard. Position the straw so its center is directly above the intersection of the N-S and E-W lines. Insert the toothpick through the straw into the cardboard at this intersection as shown in Diagram 102.

b. Take your shadow device outside during the morning and

using the compass, orient the compass card so its N is pointing North.

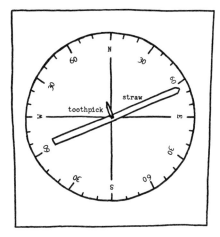

Diagram 102

c. Observe the shadow cast by the toothpick. Measure its length and record it on the chart (Diagram 103). Rotate the straw so it coincides with the shadow. Measure and record the angle formed between the shadow and the N-S line. Place an x on the compass card at the end of the shadow.

Time	Shadow Length	Shadow Angle in Degrees

Diagram 103

d. Return the compass card to the same position each hour for five or six hours during the day and repeat step c each time.

e. Construct a graph showing the time the measurements were made on one axis and the length of the shadow on the other axis.

f. Describe what happened to the length of the shadow as you approached noon. What happened to the shadow after noon passed?

g. What causes the shadow of the toothpick?

h. What happened to the shadow angle as the shadow got shorter? About how many degrees did the shadow move each hour?

i. In what compass direction did the shadow appear to move?

j. What can you say about the apparent direction and rate of the sun's motion from the toothpick's shadow?

k. How would you expect a graph to look if you had done this activity six months from now?

VOCABULARY

disc	orient
intersection	line graph

MATERIALS

compass	graph paper
compass card	ruler
cardboard	toothpick
pencil	straw

PROCEDURAL INFORMATION

It is important that the compass card be in the same position each time a measurement is made. Aligning it with the edge of a sidewalk or other fixed object would be helpful.

It may be necessary for each reading to be made by a different class. The data could then be accumulated for an entire day and shared among the classes. If so, care should be exercised in being consistent in the measuring process. Another alternative would be for individual students to gather data on their own.

The data collected should indicate that there is a constant change in the position of the sun in relation to the toothpick. Previous knowledge of the earth's motion should lead to the inference that this motion of the sun is an apparent motion.

It would be possible to modify the objective and procedures slightly for this experience to include the measuring of solar time. This would require using the toothpick and its shadow to simulate a sun dial. It could be done by adding to step c the marking of the end of the shadow on the card and recording the time.

EVALUATION

Given a graph showing the length of a shadow at three consecutive one hour intervals, predict whether the readings were taken in the morning or afternoon and what the shadow length would be for the next two hourly readings.

Experience 6: MEASURING THE MOON'S MOTION

OBJECTIVE

Measure the apparent motion of the moon over a period of several days in relation to the sun.

PROCEDURE

a. Using a newspaper, calendar, or other reference, identify the dates upon which both the sun and moon are visible during the morning. From this information, select a week or series of several days during which to do this experience.

b. Make a sketch of a stick stuck in the ground, the position of its shadow and the position of the sun. What can you say about the direction the shadow points compared to the direction of the sun?

c. Using the compass card from Experience 5, go outdoors on

the first morning that both the sun and moon are visible and place the card so that a shadow of the toothpick is visible. You should also be able to see both the sun and moon from this position. Remember never to look directly at the sun and to place the compass card in the same position each day.

d. Rotate the straw so that it coincides with the shadow of the toothpick. Record the position of the shadow from the compass card on the chart (Diagram 104) with the other data.

Date	Time	Shadow Position	Moon's Position	Angle Between Moon and Sun

Diagram 104

e. Rotate the straw so it points to the horizon directly below the moon. Observe the position of the straw and record this as the moon's position on the chart.

f. Subtract to find the angle between these two lines and record.

g. Repeat steps c and d for several days in a row. Try to make your measurements at the same time each day.

h. Construct a graph showing the size of the angle for each of the different days observations were made.

i. What was the average movement in degrees? Did the moon and sun move closer together or farther apart?

j. From the activity, can you tell whether it was the sun or the moon which was moving? Explain.

k. Do you think the motion of the earth makes any difference on the motion of the moon? Explain.

VOCABULARY

    linear relationship
    apparent motion

MATERIALS

    compass card (from Experience 5)
    graph paper

PROCEDURAL INFORMATION

The students will find that the moon and sun are both visible between the full and new moon phases. Again, it is important that the compass card be placed in the same position each day regardless of differences in position of the sun or moon.

The moon will be above the horizon and will not cause the toothpick to cast a shadow. To determine its position, the student will need to rotate the straw until it forms an imaginary line which points directly under the moon.

In the explanations of the motions, it will be evident that the moon and sun appear to be moving in opposite directions. The additional motion of the earth accounts for the rapid apparent motion of the moon away from the sun. Reference back to previous experiences will be helpful in this discussion.

EVALUATION

Given the graph (Diagram 105) describe whether it accurately shows the relationship between the motion of the sun and moon.

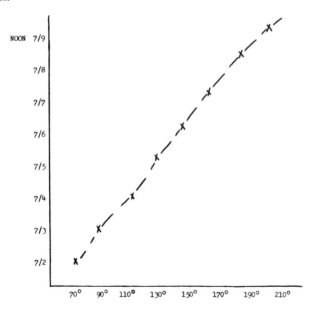

Diagram 105

Experience 7: PREDICTING MOON PHASES

OBJECTIVE

Observe and record the apparent changes in the shape of the moon, name these phases, and predict the phase of the moon when given the relative position of the sun, moon, and earth.

PROCEDURE

a. Observe the moon each night for one month and record your observations on the chart (Diagram 106).

Time	Date	Moon's Shape (diagram)	Location in Sky	Phase

Diagram 106

b. How did the moon's position in the sky change?

c. How did the shape of the moon change?

d. What might have been the cause of the times when the moon was not visible?

e. Indicate on your chart the dates during which the moon was waxing and the dates during which it was waning.

f. Identify each of the phases of the moon shown on your chart.

g. Are the phases of the moon apparent changes in its shape or real changes? Explain.

h. Darken the room except for the light from an overhead projector. While standing about three meters from the projector in its light, hold a tennis ball so the lighted side faces you. If the ball represents the moon, which phase would this be?

i. Assume that you are the earth and rotate slowly, observing the "moon" as you do so. Note the position of the earth, moon, and sun during each of the phases.

j. Make diagrams to show the positions of the earth, moon, and sun during each of the different phases identified on your chart.

VOCABULARY

phase	gibbous
new moon	quarter moon
crescent	waxing
full moon	waning

MATERIAL

paper	tennis ball
overhead projector	

PROCEDURAL INFORMATION

The moon is a part of the environment and is of most interest to students during its crescent and full phases. The new crescent is a wishing moon and the full moon is the source of moonlight. Many students are unable to identify a difference between a new and old crescent or to identify the gibbous and new moon phases. This experience will help them to become aware of the different phases and to be able to identify them.

The observations of the moon can be done by each individual in the class or as a class project with each person making an observation. It is also possible that the observations be made once every three or four days rather than every day. In selecting a starting time, be sure not to start with the new moon since it would create problems prematurely.

When simulating the moon phases, the tennis ball should be held to the front and slightly to the right or left of the head. This position should remain relatively fixed as the student turns, changing it only when the light does not strike the tennis ball (Diagram 107).

During the waxing moon, the lighted side of the moon will be on the right side when viewed from the northern hemisphere. This would include the new crescent, first quarter, and new gibbous phases. The left side will be lighted during the waning moon.

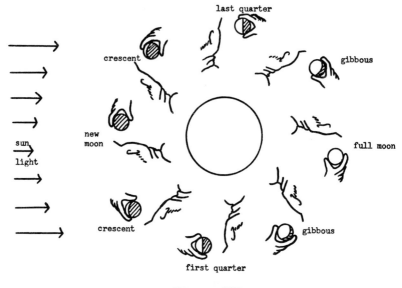

Diagram 107

EVALUATION

Fill in the circles to represent the view of the moon from the
earth for each of the positions shown (Diagram 108).

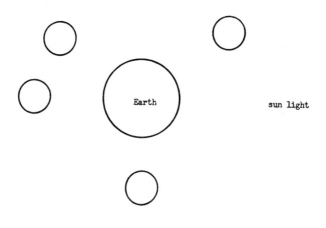

Diagram 108

Experience 8: DEMONSTRATING THE MOON RISE SCHEDULE

OBJECTIVE

Demonstrate, using a model, that the moon appears to rise later each day at a given location on the earth.

PROCEDURE

a. Draw a circle on the floor using a 3 m. length of string and a piece of chalk.

b. What is the diameter of the circle? What is its circumference?

c. The circle represents the moon's orbit. If the moon makes one revolution around the earth in approximately 28 days, how far would your "moon" have to move on the circle to represent one day?

d. Divide the circle into 28 equal parts to show the moon's daily motion.

e. If you represent the earth and stand in the center of the moon's orbit, which direction would you turn to show the earth's rotation? As you do this, have a friend represent the moon. During each rotation of the earth how far does your moon friend move? In what direction?

f. Mark a spot on the floor to indicate the first time each day when you are able to see the moon. Does the moon rise at the same time each day? Explain.

VOCABULARY

circumference	revolution
diameter	rotation

MATERIALS

string, 3 m. long   chalk
            metric tape measure

PROCEDURAL INFORMATION

When drawing the circle, the string is one-half of the diameter so the diameter of the circle would be 6 m. To determine the circumference, the student can use either C=2πr (2 x 3.14 x 3) or C=πd (3.14 x 6). When determining the relative distance the moon moves in one day, it will be close enough if the circumference is rounded off to the nearest whole number.

The moon and the earth both move in a counterclockwise direction. As the moon rotates around the earth, it becomes visible on the horizon slightly later each day, since the horizon must revolve an additional 13° each day to "catch up" with the moon. This additional distance is equivalent to about 52 minutes of time each day.

EVALUATION

Describe or show by a diagram that the moon rises slightly later each day.

Experience 9: DESCRIBING EARTH PHASES

OBJECTIVE

Demonstrate and describe the phases of the earth as seen from the moon.

PROCEDURE

a. Place a globe on a table in the center of a darkened room.

b. Use a projector to represent the sun and direct its light on the globe.

c. Starting between the sun and globe, walk slowly counterclockwise around the earth, observing the earth as you walk.

d. Where would you have to be in the solar system to observe the earth in this manner?

e. Again walk around the earth, stopping at each of the eight positions shown. Make a diagram of the lighted portion of the earth from each of these positions.

f. Compare the phases of the moon with your earth phases. See Diagram 109.

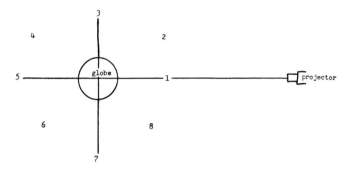

Diagram 109

VOCABULARY

full earth   phase
quarter earth   crescent earth
            gibbous earth

MATERIALS

globe   paper
slide projector   pencils

PROCEDURAL INFORMATION

Previous experiences have involved the students in observing the moon's phases from the earth. This activity demonstrates that other bodies in the solar system may have phases if the observer is in a position to view them. In this case, the student takes the position of a man on the moon. The sequence of the phases is similar to that of the moon's phases.

Some students may want to see if they can devise a model that would demonstrate phases on other bodies in the solar system. They would identify the body they are viewing and the body they are on. They could then attempt to walk through the motions of these bodies to verify their inferences.

EVALUATION

Given a series of eight circles, shade them in to show the earth phases in the sequence in which they occur and identify six phases by name.

Index